E.A Hermann

Steam shovels and steam shovel work

E.A Hermann

Steam shovels and steam shovel work

ISBN/EAN: 9783743467163

Manufactured in Europe, USA, Canada, Australia, Japa

Cover: Foto ©berggeist007 / pixelio.de

Manufactured and distributed by brebook publishing software
(www.brebook.com)

E.A Hermann

Steam shovels and steam shovel work

Steam Shovels

—AND—

STEAM SHOVEL WORK.

By E. A. HERMANN, M. Am. Soc. C. E.

1894.

ENGINEERING NEWS PUBLISHING CO.,

NEW YORK.

CONTENTS.

	Pages.
PART 1.—Steam Shovels	1–19
" 2.—Steam Shovel Work	19–41
" 3.—Disposition of Material	41–55
" 4.—Cost of Steam Shovel Work	55–57

INDEX.

Ballast, plowing	48
Blasting	39, 52
Brine, sprinkling earth	52
Cars, dump	19, 41, 47
Flat	42
Loading	19
Unloading	42, 47
Cost of work	55
Cuts	28, 36, 39
Time for	17
Widening	19
Explosives	39, 52
Fills, trestles for	47
Grades, construction track	34
Cutting down	28
Grading	25
Gravel train	42, 45, 50
Engines for	50
Unloading	48
Leveling	53
Loading cars	19
Gangs for	21, 22, 23
Operating, men for	18
Plow, Barnhart	43
Gravel	42
Plowing, cable for	50
Gravel train	48
Hauling engine for	51
Winter, brine for	52
Railways, construction	33
Reducing grades	28
Widening cuts	19
Railway work	18, 28, 33
Rapid unloader	51
Spreaders	53
Steam shovels, Barnhart	
Boilers	6
Bucyrus	9
Clement	4
Daily capacity	10
Description	41
Giant	5
Invention of	12
Little Giant	1
Industrial Works	12
Machinery of	10
Marion S. S. & Dredge Co.	5
Number of men	6
Operation	18
Osgood	16
Otis-Chapman	2
Repairs	14
Souther's	19
Thompson	14
Toledo F. & M. Co.	4
Types	8
Victor	3
Vulcan Iron Works	8
Tools	12
Track, arrangement of	16, 18
Narrow gage	10
Trains, dirt, handling	19, 42, 45, 48, 47
Trestles for fills	50
Widening cuts	47
	19

STEAM SHOVELS AND STEAM SHOVEL WORK.*

By E. A. HERMANN, M. Am. Soc. C. E.

Part I.—Steam Shovels.

The following article originated in a short paper which was read before a local society of civil engineers, and there were so many requests made for this paper and the illustrations presented with it that the author was led to believe that there was a demand for such information. Believing that a better understanding of the capabilities of these machines will serve a useful purpose in economizing money, time and labor in the execution of work to which they are adapted, the author presents in this article the information learned by a long practical experience in this special class of work. Descriptions of the various steam shovels can readily be found in the trade catalogues of the different manufacturers, but very little has been published on the manner of using them in the execution of different classes of work, and the disposition of the excavated material after it has been loaded on cars or wagons. This part of the subject will receive most attention, and although much of it may seem very elementary to those who have had an extended experience in operating steam shovels, it may be entirely new to the much larger number who have had few or no opportunities for doing work of this kind. It has been the aim of the author to condense the reading matter as much as possible, making it a point to use many illustrations in place of lengthy explanations, thus presenting the subject more clearly than by extended descriptions.

The steam shovel, or steam excavator, is a modified form of dredge adapted for excavating material on dry land. It was designed and patented by a Mr. Otis, about 1840, and like most new inventions the first machine built was a very clumsy affair.

* Copyright by Engineering News Publishing Co., 1894.

STEAM SHOVELS AND STEAM SHOVEL WORK.

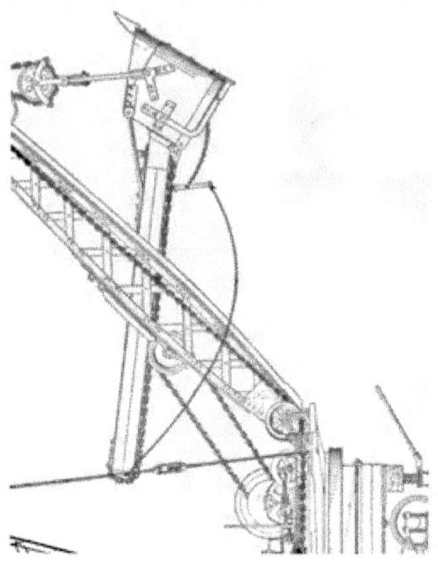

but even in this crude state it possessed many advantages for removing large masses of material. Its merits were recognized in its earliest stages, and with increased experience in its operation improvements were soon made which rendered it almost indispensable on all works requiring large quantities of excavation.

It was not until 1865, however, that the machine came into general use. About this time the largely increased railway construction created an active demand for the steam shovel, which demand was quickly supplied by several manufacturers, whose machines vary in distinctive designs of various parts, but the principles of operation are essentially the same in them all.

Types of Steam Shovels.—There are three types of steam shovels: First; machines mounted on trucks of standard gage, transported from place to place in freight trains (or propelled by their own power), and intended for railway work only. Second; machines mounted on wheels of other than standard gage, transported in sections by boat or wagon, or loaded complete on flat cars, and intended for both railway and other work. Third; machines mounted on wheels fitted for transportation over common roads, propelled by their own power, and intended for railway and other work.

The first machines built were of the second type. As now constructed they are mounted on a wide wooden frame or car body, supported by four small wheels of 7 ft. to 8 ft. gage, thus placing the machinery close to the ground, with a wide base of support. In transporting this machine from one place to another, not on the line of a railway, it is necessary to take it apart, forward the sections and put them together again at the site of the new work. The machine is built with a view to rapid dismantling and re-erection, and for work requiring a large machine for economical excavation, located in hilly country not yet made accessible by rail, or requiring transportation by boat, it is the machine most generally used. Its ready adaptability to all kinds of work in any location has made it the favorite machine with many general contractors whose work includes large contracts for railway and other excavation. For transportation by rail this machine is run onto an ordinary flat car, only the crane being detached and loaded on a separate car. With this manner of shipment the machine can be made ready for railway work very quickly, but for exclusive railway work

FIG 2.—THOMPSON STEAM SHOVEL; B Steam Shovel & Dredge Co., South Milw. Wis.

a machine of a later design has come into use and is now generally preferred for this class of work.

This is the machine of the first type, resting on a wooden or iron car body, supported on trucks of standard gage, with an iron or steel crane from 18 to 26 ft. high over the track when in working order, and which can be lowered to 14 ft. to permit shipment through tunnels and under low overhead bridges.

Machines of the third type are generally of smaller capacity than the others; they have come into general use only within the past few years, but are now multiplying rapidly in numbers as their utility for nearly all kinds of work is better appreciated. They are especially adapted to smaller jobs and work not readily accessible by rail, but where common roads are available.

These three types are shown in Figs. 1 to 9, representing the machines of seven of the principal manufacturers.

Steam shovels will excavate any kind of material except solid rock, and they will load rock if it has been broken up by explosives into pieces of not more than 3-4 cu. yd. in size. The materials excavated by them are mostly sand, loose gravel, all kinds of clay, cemented gravel, hardpan, clays mixed with bowlders and other small stones, ore, phosphate rock, loose rock and thin seams of slate, shale or sandstone.

These machines are used for excavating material, loading it on cars or wagons for ballasting tracks; for filling trestles, streets, roads, dams, lots and new city additions; for widening embankments for double track, side tracks, yards, shops and station grounds; for cutting down street, road and railway grades; grading lots and new city additions, railway yards, shop and station grounds; widening cuts, removing land slides, stripping coal fields, ore beds and stone quarries; digging canals and drainage ditches, loading clays for brick yards, etc.

Construction of Steam Shovels.—The general plan of construction of the machines, shown in Figs. 1 to 9, is essentially the same in all, and consists of a strong frame, mounted on wheels, forming the base to which all working parts are attached. The boiler and machinery are placed near the rear end of the frame, and the mast, or post, and crane at the front end. The crane is made in two pieces connected only at the top or point, and at the foot of the mast. Between these pieces, serving as guides, is the dipper handle, carrying at its farther end the dip-

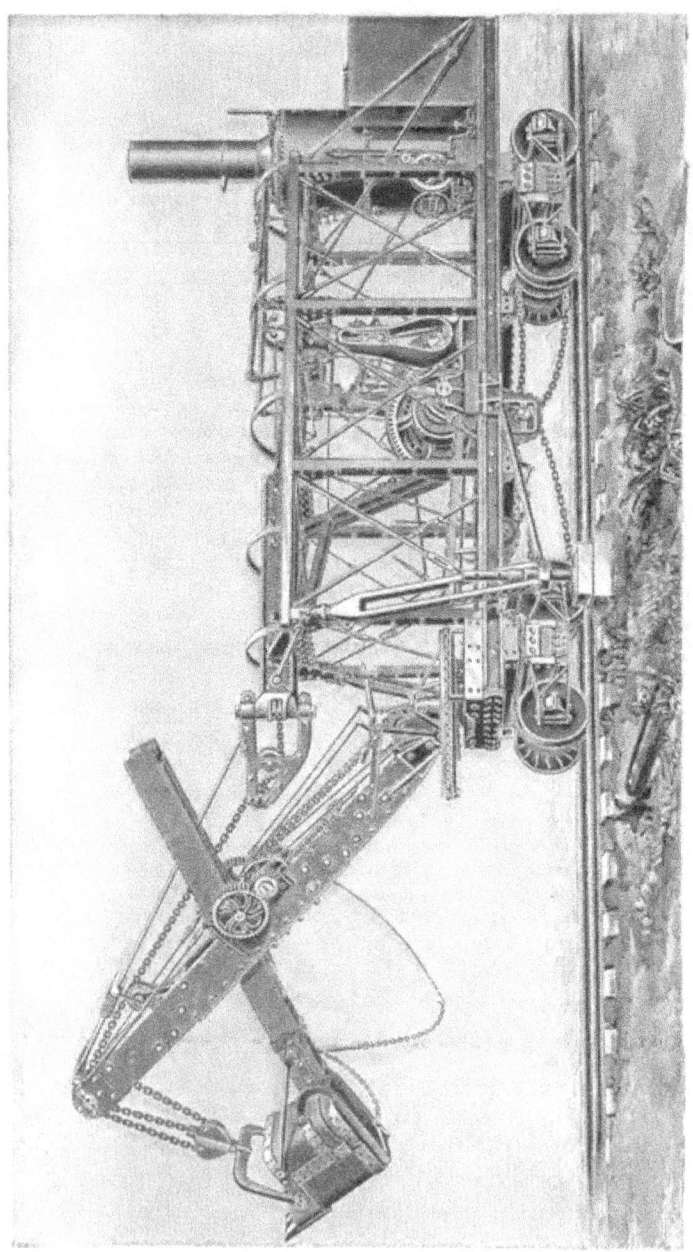

FIG. 3.—BARNHART STEAM SHOVEL; Marion Steam Shovel Co., Marion, O.

per or scoop. To the top of the post (or to the foot in some machines) the swinging circle is secured.

The most used, and hence the most important part of the machinery of the steam shovel is the gearing imparting motion to the hoisting drum, actuating the chains by which the dipper is raised and lowered. It is in almost constant use, and is often subjected to severe shocks in hard digging. Of all parts of the machinery it is the most likely to break or rapidly wear out. Naturally it has received the most attention of any part of the steam shovel in all efforts to improve the design, strength and durability of the machine. There are a number of different gears in use, and essentially they are either friction clutches or positive gearing. The use of the former subjects the machinery and crane to less severe shocks, and can be thrown in and out of gear more rapidly, but it wears out quicker, often causes delay by heating, and requires frequent repairs. Positive gearing exposes the machinery and crane to more severe shocks in hard digging, and must be started slower, especially in hard material, but while these machines are a little slower than those operated with friction clutches, they are less subject to the expense of repairs and delay due to the disarrangement of the hoisting gear, so that their total output of material about equals, and sometimes exceeds, the quicker moving friction gear machine.

The mechanism for thrusting the dipper into the bank is attached to the crane, and the forms most generally used are as follows:

1. A chain, one end of which is attached to the rear end of the dipper handle, and the other end wound around a drum receiving its motion by an endless chain passing over a sprocket wheel connected to the axle of the sprocket wheel at the top of the mast, over which the hoisting chain passes, thereby revolving both wheels. This drum is thrown into gear by a friction clutch, and its motion regulated by the cranesman's lever and footbrake.

2. A rack on the dipper handle operated by a pinion attached to a shaft revolved and regulated as above described.

3. A small double cylinder engine operating either a pinion and rack as above described, or revolving a drum with a chain attached to it, and the rear end of the dipper handle as described in the first case.

4. A long steam cylinder attached to the dipper handle, whose

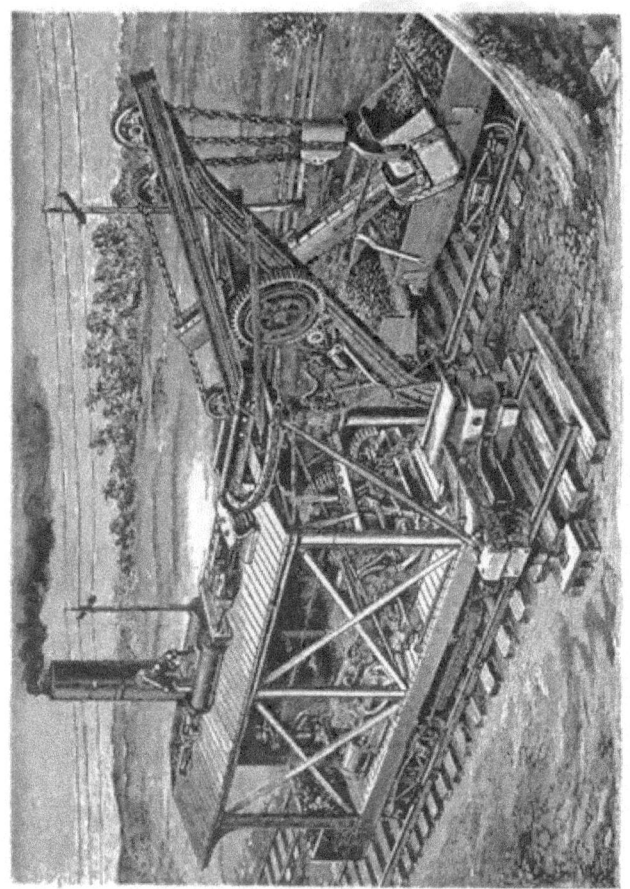

FIG. 5.—VICTOR STEAM SHOVEL; Toledo Foundry & Machine Co., Toledo, O.

piston rod is connected to the dipper, extending or withdrawing it as desired.

The thrusting mechanism used in the last two cases imparts a rapid, steady and powerful motion, but the extra engines or steam cylinder and their connecting steam pipes involve a complication which often more than balances their advantages.

Swinging the crane in a horizontal direction is generally accomplished in one of the following three ways:

1. A chain passing around the swinging circle attached to the post, and wound around drums connected to the engine by positive gearing or friction clutches.

2. A wire rope passing round the swinging circle and connected to the piston rods operated by two long steam cylinders.

3. A chain passing round the swinging circle and wound around a drum connected to a small reversible engine.

The mechanisms used in the last two cases have the same advantages, but also suffer from the same objections urged against employing small engines or a steam cylinder for thrusting the dipper into the bank.

The engines are either of the upright type with a single steam cylinder, or of the horizontal type, with double horizontal steam cylinders. The size of the cylinders varies for machines of different capacities, ranging from 8 by 10 ins. to 10 by 12 ins. for the upright engines, and 6 by 8 ins. to 13 by 16 ins. for the horizontal engines.

The upright type of boiler with submerged flues is usually preferred, as it occupies only a small space. Horizontal boilers of the locomotive type are used in a few machines, and are more economical in the use of fuel, but occupy too large a floor space. Forced draft is used in both types of boilers, and they are generally worked to the limit of their capacity. The usual working pressure is 90 lbs. per sq. in. The safety valve is generally set to blow off at 120 lbs. per sq. in. The boiler is supplied with water either from an upright circular sheet iron tank located in a corner of the machine, behind the boiler, or from a sheet iron box tank hung under the floor. These tanks usually hold about 1,000 gallons of water, enough to run the machine half a day. The water is obtained by a pump or siphon from the tender of a locomotive on railway work, or is hauled to the machine by wagon on other work.

In some machines the frame or car body is made of wood,

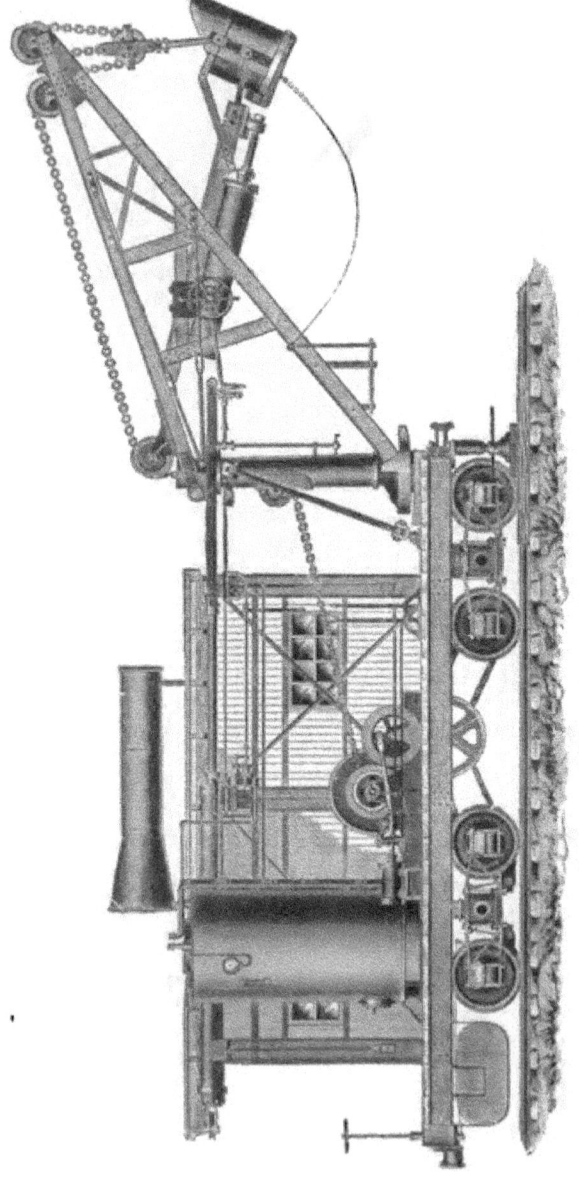

FIG. 6.—CLEMENT STEAM SHOVEL; Industrial Works, Bay City, Mich.

generally oak, often incased with heavy plate iron. In others it is constructed of iron or steel I-beams and channels. In all machines it is strongly built and braced with a view to sustain the weight of the working parts and to resist the shocks to which it is subjected. The floor is usually of 3-in. oak plank.

The mast or post is made of cast or wrought iron, strongly braced and guyed to the frame. It is the pivot about which the crane swings, and easy working in its bearings is of great importance for the rapid and economical operation of the machine. In order to prevent breakage or delay it should never be permitted to wabble by neglecting to promptly tighten its braces and guys in case they should work loose. The post should always stand vertical, or practically so, to insure the horizontal motion of the crane and avoid unnecessary straining of the swinging gear. For this reason the machine should be set practically level before beginning operations; and using a small mason's level is better than trusting to the eye, when blocking under the track and adjusting the jack screws for this purpose.

The crane is secured to the post, and is made of wood, iron or steel, strongly and compactly built to resist the shocks to which it is often subjected. It is from 14 to 20 ft. high above the track or ground, varying with machines of different sizes and manufacture, and swings horizontally through an angle of 180 to 240 degrees, with a radius of 15 to 20 ft. In some machines it must be detached from the post for shipment, in others (mostly those made for railway use exclusively) it can be lowered to a height of 14 ft. above the track, thereby permitting shipment without detaching from the post.

The dipper, scoop, or bucket is made of iron or steel, shaped somewhat like a coal scuttle. Its cutting edge is protected by four teeth made of steel or steel pointed. These teeth are easily removed for sharpening or replacement. Dippers vary in size from 1-2 cu. yd. to 2 1-2 cu. yds. capacity. They also vary somewhat in shape, according to the material to be excavated, though no special provision is made for this unless there are very large quantities of the same kind of material to be removed; or for machines working in a certain class of material only, like ore loaders. For general work in all kinds of materials the dipper is seldom changed.

For soft, tenacious material, likely to adhere to the inner sides of the dipper, and not drop out promptly when the bottom door is opened for unloading, the dipper is shaped as shown in Fig. 10, with a larger bottom than mouth. In hard, or dry soft material the section shows parallel sides, as in Fig. 11. For general use the bottom of the dipper should be slightly larger than the mouth, as most materials contain more or less moisture

FIG. 7.—GIANT STEAM SHOVEL; Vulcan Iron Works Co., Toledo, O.

which is likely to produce a partial clogging of the dipper by material sticking to the inner sides, especially between the teeth, necessitating frequent cleaning out whenever the machine is stopped while preparing to move forward, and sometimes oftener. For ordinary clay, cemented gravel, and hard dry

FIG. 8.—LITTLE GIANT STEAM SHOVEL; Vulcan Iron Works Co., Toledo, O.

materials, a dipper with a wide and shallow mouth, as shown in plan in Fig. 12, is preferred to the one shown in Fig. 13, which latter is better adapted for loose gravel, sand and other soft dry materials where a deep cut can easily be made. For hardpan, shale, loose rock and similar materials, ample strength of teeth and dipper is of great importance than its shape.

To prevent tenacious material from sticking to the inner sides of the dipper, and to allow it to drop out freely when the bottom door is opened, it is often good economy to place a barrel of water near the head of the machine from which a bucketful can be taken and thrown into the dipper just before each cut. The water acts as a lubricant and causes the material to drop out more readily. For cleaning the dipper, the tool shown in Fig. 14 is used.

The chains have links of three-quarters-inch to one-inch diameter, and are made of iron, sometimes of steel. Their constant use necessarily subjects them to great wear, and as they are also often exposed to severe shocks (especially the hoisting chain) they must be made of the very best material and in the most

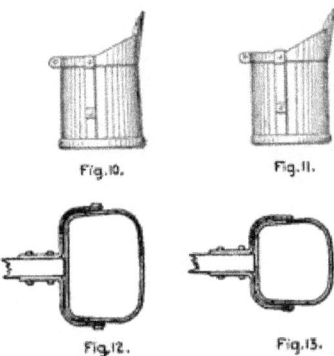

Figs. 10 to 13.—Buckets for Steam Shovel.

careful manner. At present iron chains are preferred to those made of steel; they are more durable, and less likely to break under severe shocks. Steel chains have suffered in reputation through rapid wear and frequent breakages occurring within the last few years, but with increased experience in their manufacture and use they will undoubtedly be improved, and eventually take the lead over iron chains.

The propelling mechanism consists of an endless chain connecting one or more axles of the truck or supporting wheels with the shaft of the hoisting drum by means of friction clutches or positive gearing. The usual speed is five to six miles per hour.

Steam shovels of seven of the most prominent manufacturers

14   STEAM SHOVELS AND STEAM SHOVEL WORK.

FIG. 9.—OTIS-CHAPMAN STEAM SHOVEL; John Souther & Co., Boston, Mass.

STEAM SHOVELS AND STEAM SHOVEL WORK. 15

are shown in Figs. 1 to 9, and the general particulars of each are given in condensed form in Table I. In each case the boiler is upright.

TABLE I.—GENERAL DESCRIPTION OF THE IMPORTANT PARTS OF THE MOST PROMINENT MAKES OF STEAM SHOVELS.

Fig.	Shovel	Frame Material	Frame Size, ft.	Running gear	Gage, ft. in.	B'l'r	Engine	Cylinder ins.	H'st'g gear
1..	Osgood	Wood	10 × 34	2 trucks	4 8½	Vert.	Hor.	Two 10 × 12	
	"	"	10 × 30	"	" "	"	"	" 8½ × 10	
	"	"	10 × 25	"	" "	"	"	" 7 × 10	
2..	Thompson	I be'm and chan- nels	10 × 32	"	" "	"	"	" 10 × 14	Friction clutch
			10 × 30	"	" "	"	"	" " × "	
			10 × 28	"	" "	"	"	" 8 × 12	
			10 × 26	"	" "	"	"	" 8 × 10	
			10 × 24	"	" "	"	"	" 6 × 8	
3..	Barnhart	"	10 × 28	"	" "	"	Vert.	" 9 × 10	
	"	"	10 × 26	"	" "	"	Hor. One	8 × 10	
	"	"	10 × 24	"	" "	"	Vert.	" 8 × 10	
	"	"	10 × 22	"	" "	"	"	6 × 8	
5..	Victor	"	10 × 30	"	" "	"	Hor. Two	8 × 10	
6..	Clement	"	10 × 30	"	" "	"	" "	8 × 10	Pos'ive
7..	Giant	"	10 × 35A	"	" "	"	" "	18 × 16	
	"	"	10 × 35	"	" "	"	" "	8 × 12	Friction clutch
	"	"	10 × 30	"	" "	"	" "	7 × 11	
8..	Little Giant	"	7 × 23	4 r'd wh.	8 0	"	" "	7 × 11	
			6 × 23	"	8 0	"	" "	6 × 8	
9..	Otis-Ch'pm'n	Wood	10 × 22	4 fl'ge. wh	7 10	Vert.	One 10	× 12	Posi-
	"	"	10 × 18	"	7 10	"	"	8 × 10	tive

Fig.	Thrusting Mechanism.	Swinging Mechanism.
1	Reversible engines, 2 steam cylinders each 6 × 8 ins.	Chains attached to circle geared to hoisting drum
2	Do., do 5 × 6 ins.	
3	Rack on dipper handle actuated by friction clutch geared to hoisting drum.	
5	Reversible engine, 2 steam cyls. 6 × 8 ins.	Wire ropes attached to circle and pist'n rods in long st'm cyl.
6	Long st'm cyl., piston rod att'ch'd to dipper	Reversible engine, 2 steam cyl- inders 5 × 6 ins.; except A, cylinders 7 × 9 ins.
7	Reversible engine, 2 steam cyls. 5 × 6 ins.	
8		
9	Chains on dipper handle actuated by fric- tion clutch geared to hoisting drum.	Chains attached to circle geared to hoisting drum

Fig.	Post material.	Crane. Material.	H'ght ab've gr'nd or track. Work- ing or- der, ft.	Ship- ping or- der, ft.	Radius, ft.	Swing- ing angle, deg.	Capacity of dipper, cu. yds.	W'ht. tons.
1	Wt. iron A frame	Wt. iron " "	28 24 20	14 14 14	24 24 20	240 240 240	2 1¾ 1	40 30 20
2	Cast iron	" " "	23 18 18 16	14 14 14 14	20 18 18 12	200 200 200 200	2½ 1⅝ 1¼ ¾	45 40 30 20
3	Wt. iron	Wood	26 24 20 18	14 14 14 14	20 20 18 18	200 200 200 200	1½ 1 ¾ ½	37 26 16 12
5	Hollow wt. ir.	Wt. iron.	19	14	20	200	2	40
6	Cast iron	"	20	14	20	200	2	40
7	Cast steel	Steel	20	14	19	200	2½	70
	Cast iron	"	20 18	14 14	19 17	200 200	1⅝ 1¼	45 30
8	"	"	16 15	Detach'd	15 15	185 185	1¼ ¾	20 18
9	"	Wood	20 16	" "	20 18	200 200	2½ 1½	26 15

Makers: 1 (Osgood): Osgood Dredge Co., Albany, N. Y. 2 (Thompson): Becy- rus Steam Shovel & Dredge Co., Bucyrus, O. 3, 4 (Barnhart): Marion Steam Shovel Ca., Marion, O. 5 (Victor): Toledo Foundry & Machine Co., Toledo, O. 6 (Clement) Industrial Works, Bay City, Mich. 7 (Giant) and 8 (Little Giant): Vulcan Iron Works Co., Toledo, O. 9 (Otis-Chapman): John Souther & Co., Boston, Mass.

Operation of Steam Shovels.—All movements of the steam shovel are controlled by two men, the engineman and the cranesman. The former is stationed near the engine, the latter on a small platform attached to the crane. The engineman directs the movements for raising and lowering the dipper, swinging it into position for unloading, and moving the machine forward or backward. The cranesman regulates the depth of the cut made by the dipper, releases it from the bank when full or near the top of the crane, and pulls the spring latch of the bottom door of the dipper when in position for unloading, thereby dumping its contents.

These motions are shown in Figs. 15 and 16. Beginning with the dipper in the position shown at A, Fig. 15, the engineman throws the hoisting drum into gear, and starting the engine pulls the dipper upward, the cranesman at the same time thrusting it forward, regulating the depth of the cut so that it will not stop the engine or tip up the rear end of the machine. When the dipper has reached the position B, near the top of the crane,

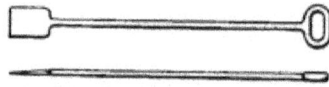

Fig. 14.—Spade for Cleaning Buckets.

the engineman throws the hoisting drum out of gear, and holds it in position with a foot brake; at the same time the cranesman by easing his foot brake, allows the dipper to fall back to the position C. The engineman then swings the dipper over the car or wagon, as shown in Fig. 16, when the cranesman pulls the latch rope, thereby opening the bottom door of the dipper and dropping the contents. The engineman then swings the crane back again to the next cut, at the same time releasing his foot brake on the hoisting drum until the dipper has fallen to a point near the ground, as at D, Fig. 15, where he holds it for an instant with the foot brake, then drops it by releasing the brake, while the cranesman (during this slight drop) regulates the length of the radius of the dipper handle by releasing his foot brake so as to bring the dipper into the position A again, and adjoining the last cut. While the dipper is being lowered, the bottom door closes and latches itself by its own weight, when all is ready again for another cut.

These motions are very simple when taken separately, but when performed together by two different men, experience and quickness in both are required to carry on the work rapidly

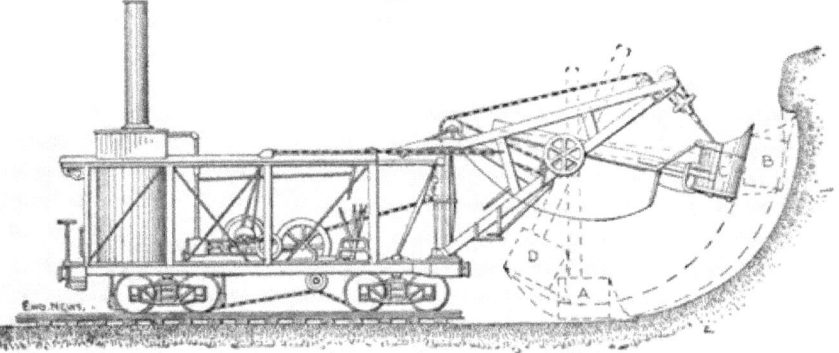

Fig. 15.—Showing Series of Operations for Excavating.

and harmoniously, without breakages or delays. In loose gravel one cut can be made in a half to three-quarters of a minute; in hard materials one and a half to two minutes, seldom more.

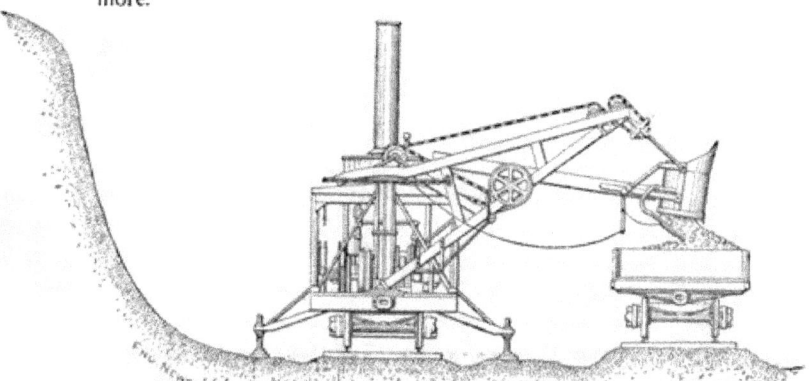

Fig. 16.—Loading Earth from Steam Shovel Onto Cars.

After all material within reach of the dipper has been removed, an unoccupied section of track (generally about 4 ft. long) at the rear of the steam shovel is attached to the dipper by a chain and dragged around the machine to the front (by swinging the dipper horizontally) and there placed in position in line with the sections of track under the machine. The screws at the ends of the jack arm (a horizontal bar at the front end of the machine used for steadying it when cuts are taken at right angles to the steam shovel) are then released, and the machine moved forward three or four feet by throwing the propelling gear into motion. After placing the jack screws into their new position, and tightening them, and blocking the supporting wheels of the steam shovel, the machine is ready for another series of cuts.

The regular employees for operating a steam shovel are the engineman, cranesman, fireman and four laborers. The latter are under the supervision of the cranesman, and their duties are to shovel forward any lumps or loose material which may roll down and lodge too close to the front of the steam shovel to be reached by the dipper, to level the surface of the ground in front of the machine, preparing it for the next section of track,

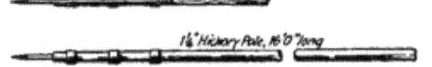

Fig. 17.—Pole for Breaking Down Edge of Excavation.

to lay these sections of track, to attend to the jack screws and blocking and to act as general utility men.

With this crew dry sand and loose gravel can readily be loaded. In harder or more tenacious materials from two to six extra men are required, depending upon the kind of material to be excavated, and also upon good management of the contractor or foreman in charge. Wet sand and fairly loose gravel requires only two extra men, whose duty is to break down the overhanging ledges or these materials which cannot be reached by the dipper, and are liable to fall when the machine has advanced, burying it or blocking the pit behind it. The implement used by these men is a pole, Fig. 17, headed by an iron point, resembling a surveyor's pole. With these poles fairly loose gravel and sand can be readily broken down, sloped at its natural angle, and fed into the pit in front of the steam shovel. In harder materials three to four extra men are usually sufficient, but in very hard or tenacious materials as many as six must be employed. These men break down overhanging material in the face of the bank which cannot be reached by the dipper, bore or drill holes for powder or dynamite when blasting becomes necessary, cut and remove trees, etc.

On all but very small pieces of railway work there are also employed a blacksmith and helper, and two to five car repairers. The blacksmith's work consists mostly of repairs about the cars, mainly bent or broken aprons, sideboards, chains, etc. The steam shovel occupies much the smaller part of his time. His

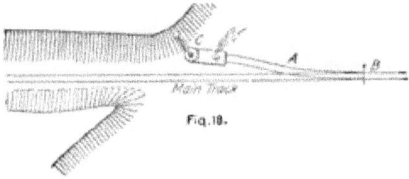
Fig. 18.

accommodation requires a small rough frame shop about 10 by 16 ft. (an old box car body is frequently used), with forge and tools. Another rough frame shed of about the same size is needed for the storage of tools, oils and supplies. The sectionmen of the respective sections are occasionally called on for the building and maintaining (or taking up) of the various side tracks required during the progress of the work.

Part II.—Steam Shovel Work.

Widening a Cut; Loading on the Main Track.—The simplest and one of the most frequent cases for the application of a steam shovel is the widening of a single track railway cut. The man-

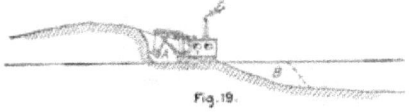
Fig. 19.

ner of doing this is shown in Fig. 18. A switch, A B, is put in the main track just beyond the end of the cut and far enough away to permit the steam shovel (when standing on the side track) to clear cars on the main track. Cars are then placed opposite it on the main track and the machine is ready for excavation.

It very frequently happens that the end of the cut joins directly on an embankment, as shown in profile, Fig. 19. In cases of this kind it would be necessary to widen the embankment for the reception of the side track, near the end of the cut, if the machine were to begin work at that point, C, Fig. 18. This is very seldom done; the usual method is to remove the section, A, Figs. 19 and 20, to B by hand labor with wheelbarrows or with teams and scrapers. The excavated material is used to widen part of the embankment near the end of the cut for the reception of the side track. Section A is made

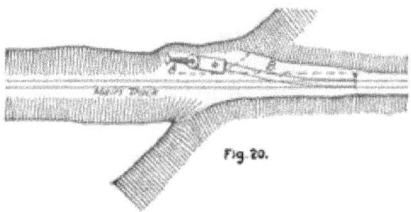

Fig. 20.

barely long enough to provide a standing place for the steam shovel and clear cars on the main track; it is seldom over 50 ft. long, and averages about 30 ft. After placing the machine in this space it is ready for work. Strings of 10 to 20 cars are then drawn along the main track, and stopped opposite the machine for loading.

When the machine has reached the end of the switch, it advances on short sections of track, generally 4 ft. long, which are

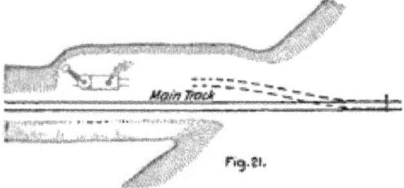

Fig. 21.

placed in front of it, and again taken from its rear when it has moved forward one section of track more than its own length. When no more cuts are to be made for still further widening, the switch is taken up again and the machine advances on its own track sections, Fig. 21. When other cuts are to follow, however, a loading track is needed for the next cut; the side track is then extended for this purpose at convenient intervals, generally about 300 ft. at a time though often after each space of a rail length (usually 30 ft.) is clear. The latter is by far the best practice, as it permits the immediate withdrawal of the machine in case of a threatened cave-in, sidehill slip, or other unforeseen danger.

After all the cars have been loaded they are taken away for unloading. Sometimes the steam shovel is left idle until the train returns, which is a very wasteful method of working, even where the haul to the dump is short, half a mile to two miles. Two engines and crews should be furnished for hauls up to ten miles; three engines and crews, or more, for longer hauls, or where the traffic on the main line is very heavy, and delays to the work trains are frequent. The material is generally utilized in filling trestles, widening embankments for side tracks, double tracks, yards, etc., thereby making two improvements at the same time.

In widening a cut it is good policy to keep the grade of the pit from 1 to 2 ft. below the surface of the subgrade of the main

Fig. 22.

track, as shown in Fig. 22, thereby providing for drainage of the ballast and also providing a receptacle for the spreading of loose material dropping off the cars and washing in from the surface of the cut; there is nearly always considerable of this loose material to roll or wash into the pit after the cut has been completed; and unless room is provided for it, the accumulation will soon reach the height of the track, washing mud on it, and choking the drainage, thus injuriously affecting the main track.

Widening a Cut; Loading on a Side Track Graded by Hand or Steam.—The delays in loading on the main track of a railway in operation, due to the clearing of the track for all trains, vary from one to four hours per day of ten hours, and sometimes amount to as much as seven hours, depending upon the density of the traffic on the line. The first cut in a case such as the latter is therefore necessarily an expensive one, and where the traffic is so heavy it is often cheaper to make a narrow cut for the side track, on which the steam shovel is to load, either by wagons and wheel scrapers, Fig. 23, or by hand with wheel-barrows loading back on cars, Fig. 24.

The latter plan has the great disadvantage that only one car at a time can be loaded and only few men (six to ten) can be

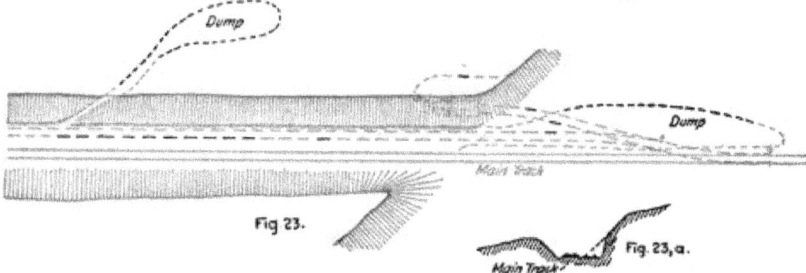

employed. Therefore this plan is never adopted where quick work is required, but is used only where ample time is available, and mostly as an early spring preliminary job, preparing the way for the operation of the steam shovel later in the season. From three to six flat or coal cars are used, enough to require a whole day for the gang of men employed to load; the material from the face of the excavation is loaded on wheelbarrows, and

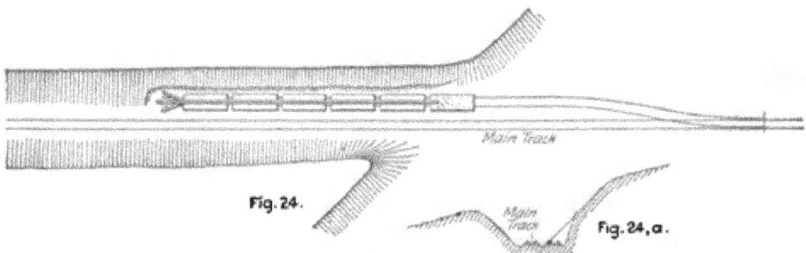

wheeled over the empty cars to the one farthest from the cut. This car is loaded first, then the one next to it, etc. At night the loaded cars are taken out of the switch by the first available freight train and hauled to the nearest yard or side track where widening of the embankment is wanted, or where the material can be otherwise used to advantage, and there unloaded by a small gang of men on the following day; the cars to be returned again the next night. Other empty cars are placed in the pit track for loading next day, by a train bound toward the pit the same night the loaded cars were taken out. The work can be carried on from either one or both ends of the cut. Coal cars should never be used if flats can possibly be obtained, as the latter can be unloaded by a gang of men one-third as large as would be necessary for unloading coal cars.

Sometimes small dump cars are used, drawn by horses or mules, and the material unloaded at the end of the cut, thereby widening the embankment for a long side track, Fig. 25. The narrow gage track, A, is laid over the ditch adjoining the main track; the material for any slight excavation that may be necessary for this track is shoveled on the slope of the cut, as at C, on the cross section. The material is then loaded on small

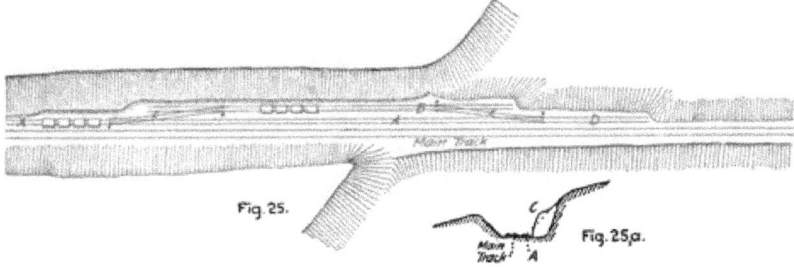

Fig. 25. Fig. 25,a.

dump cars standing on track A, and unloaded at D. The cars are returned on track B. The cross-overs, E and F, are taken up occasionally and relaid near the advancing ends of the cut and dump.

In short cuts the narrow excavation necessary for placing a side track in the cut for the steam shovel to load on is generally taken out by carts and dumped at the ends of the cut, widening the embankment for a long side track.

The plan of excavation with wagons or wheel scrapers for this side track, shown in Fig. 23, is adopted where the traffic is too heavy to permit loading on the main track; when the side track is wanted at the earliest possible time; and in cuts not over 40 ft. deep. The material is dumped at the ends of the cut until the haul becomes too long, then it is taken to the top of the cut over sidehill driveways excavated for the purpose, and unloaded at a sufficient distance from the edge of the new cut to prevent its washing back by rains.

These expedients are necessary only on railways where traffic is very heavy. On most railways (on all where the total delay does not exceed five hours per day) it is cheaper to load on the main track until the first cut has been made. This necessarily involves the delay due to running to and returning from the nearest side track to get out of the way of every main line train,

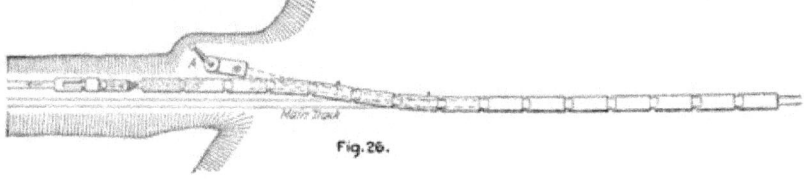

Fig. 26.

until the pit track is long enough to contain the construction train. This, however, seldom requires more than two weeks, generally only one; the excavation of all of the first cut does not often occupy more than a month, and is only a very short time compared with the whole length of time that the steam shovel is usually in operation on all but very small jobs.

After a side track has been laid in the first cut made by one of the methods described above, the steam shovel begins work at A, Fig. 26, loading cars standing on the side track, and some

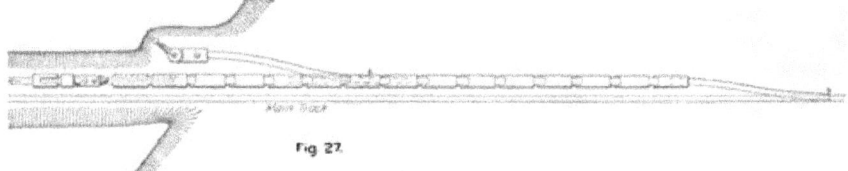

Fig. 27.

of them extending out on the main track. At first not more than ten cars should be coupled to the engine, so that the train can quickly run into the side track on the approach of a main line train, and not delay its passage. After the steam shovel has advanced a train-length, the full number of empty cars can be coupled to the engine, as they will all be on the side track while being loaded.

Where the embankment has been previously widened by the

excavated material from the cut, Fig. 27, a sufficient length to permit laying a side track long enough to hold the construction train, the full number of cars can be used at once, a great advantage in keeping the steam shovel at work without interruption by passing trains, which is unavoidable when some of the cars extend out on the main track.

After the machine has reached the other end of the cut it is either withdrawn for other work, or placed on the other side of the main track for widening the cut on that side. The steam shovel begins at A, Fig. 28, loading cars standing on the main track; the main line traffic being carried over a temporary main track built in the excavation previously made by the steam shovel on the other side of the main track. Only a few cars at a time can be used for loading at first, unless the temporary main track has been extended toward B a sufficient length to clear the usual string of about 20 cars when the first car is being loaded.

Grading Wide Areas.—In loading gravel for ballasting, or in widening a cut for the purpose of grading yard, shop or station grounds, the usual manner of doing the work is shown in Figs. 29 to 34. After the first cut has been made by one of the methods already described the steam shovel is started in at A, Fig. 29, for the second cut. After its completion the first side track becomes available for the storage of empty and loaded cars as in Fig. 30, greatly increasing the convenience of handling the cars and preventing delays by interferences between the strings of empty and loaded cars, then the latter cannot be taken away promptly on account of passing or shortly expected trains on the main line. After the completion of the third cut, another side track is available for cars, Fig. 31, the loaded cars are then placed on the first inside track and the empty ones on the second. The former are taken away by the road crew, and on their return placed on track No. 2. The pit crew set their loaded cars on track No. 1 for the road crew, and get their empties from track No. 2. The pit track in the rear of the steam shovel is used as a repair track for cars.

After the completion of the fourth cut, Fig. 32, track No. 3 is used for a car repair and extra storage track for loads or empties, for which there may not be room in tracks 1 or 2. Enough tracks have then been built for the most efficient and economical handling of the loaded material, and if the empty

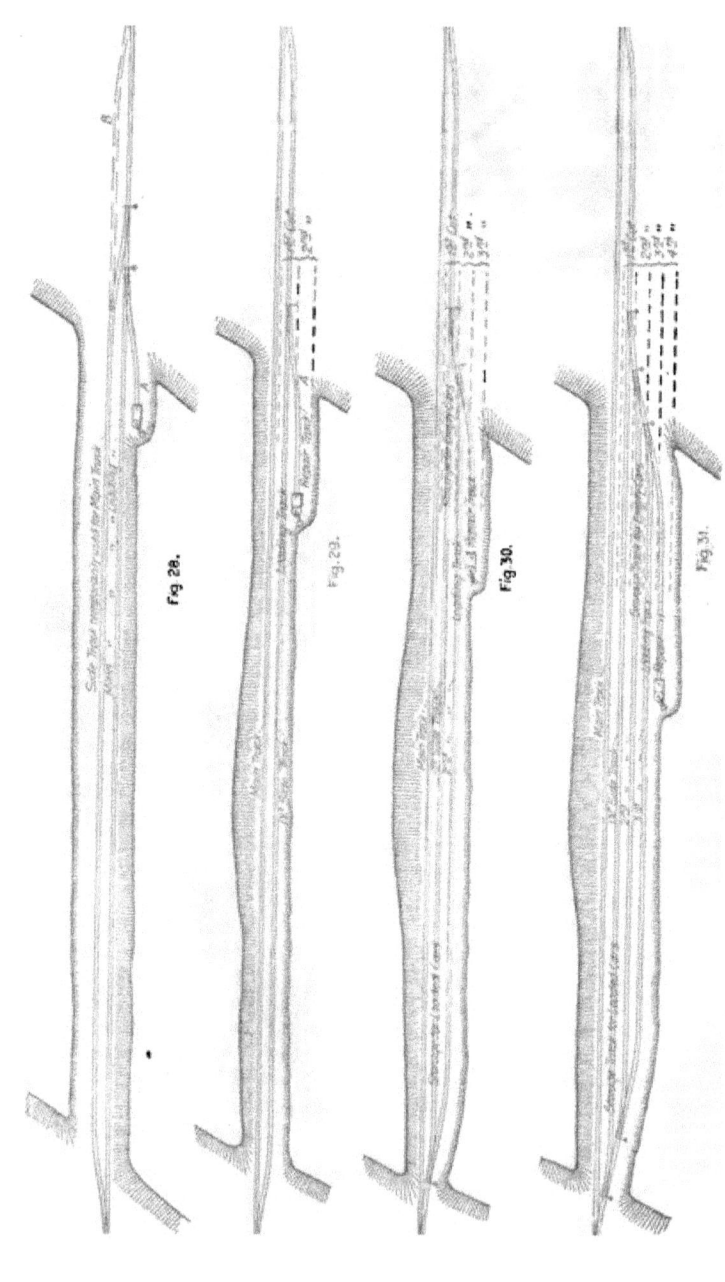

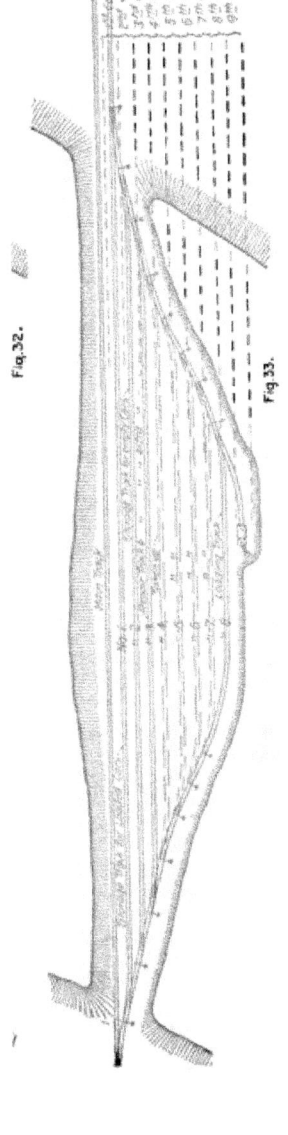

Fig. 32.

Fig. 33.

cars are promptly returned the steam shovel can be kept almost constantly at work. Each pit track, on which the steam shovel advances, becomes a side track on the completion of that cut, to be used as a loading track for the next cut up to the fourth cut, after which the loading tracks are taken up on completion of the cut for which they are used, Fig. 33, and relaid in the pit of the next cut, to be used, taken up, and relaid as before for the following cuts. In pits less than one-quarter mile in length, it is sometimes necessary to retain more of these tracks to provide ample storage space for all loaded and empty cars.

On all large pieces of work where the main line traffic is heavy it is important that the first side track from A to B, Fig. 32, shall be of sufficient length (usually about 700 ft.) to hold the engine and a full string of cars to avoid going on the main track when switching loads to C, and obtaining empties from D. If there is an embankment from A to B it can be widened with material taken from the cut, either by wagon or cars.

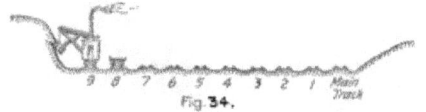
Fig. 34.

Grading by this method for yard, shop and station grounds occurs mostly near large cities where better terminal facilities must be provided for. The width of the area excavated in this manner seldom exceeds 200 ft. (eight cuts) except in old gravel pits used for furnishing material for ballasting track, which are sometimes 300 ft. (twelve cuts) or more in width.

Gravel pits and other wide areas excavated are seldom less than one-quarter mile or more than one mile in length. One-half to three-fourths of a mile is the most usual length; in exceptional cases two miles have been reached. Long and narrow pits can be worked more advantageously than short and wide ones.

Cutting Down Grades.—For cutting down grades on railways where the traffic is not too heavy to prohibit loading on the main track, the usual plan of operations is shown in Figs. 35 to 42. The machine begins work at A, Figs. 35 and 36, the beginning point of the new grade, loading cars on the main track, cutting to the line of the new grade, and moving forward on the track on the surface of the pit as long as the height of the crane permits raising the dipper high enough over the cars to open the bottom door of the dipper and discharge its contents, B, Fig. 35. This point is usually about 2 ft. below the main track. The machine must then be gradually run upward on a cribwork of wooden blocking, generally pieces of pine 6 by 12 ins. by 4 ft. long, with some longer track stringers for supporting the sections of track on top of the blocking, and some thinner pieces for attaining exact heights of blocking when needed. As the machine moves forward the dipper still continues cutting to the line of the new grade, while the machine is gradually run upward on the blocking on a grade parallel to the grade of the main track, and slightly below it, maintaining a constant height between the top of the track on the blocking and the highest point to which the dipper can be

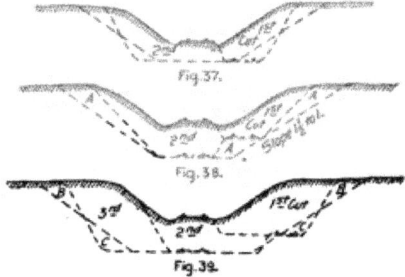

raised on the crane to insure discharging its load on the cars. When the dipper has cut as low as the length of the dipper handle will permit, C, Fig. 35, the greatest depth to which the machine will cut below the level of the main track has been reached, and as the steam shovel advances the surface of the pit will be on a grade parallel to the grade of the main track, running upward to the summit, S, then downward, and continue so until it cuts the new grade line at H, when the dipper is made to cut on this grade, while the blocking under the machine is gradually lowered as it was previously raised, until the steam shovel reaches the end of the new grade at I, when it is again on the surface of the pit.

Although the machine is gradually run upward and downward, it is always blocked level after each forward move before

30   STEAM SHOVELS AND STEAM SHOVEL WORK.

beginning work, to insure quick and easy swinging of the crane, as previously explained. Most machines will cut 5 ft. below the main track and load on a flat car with 18 ins. side boards. Some machines will cut as low as 8 ft., and they are preferred to others on railways where much work of this kind is done, as their use often avoids making an extra cut.

After the first cut has been completed, the pit track, A 1, Fig. 36, becomes the temporary main and loading track; the main track is taken up from C to H, and the steam shovel run back to C to begin the second cut, Fig. 42, excavating it in the same way as the first, and loading on the temporary main track. This track again is taken up after the second cut, the machine begins

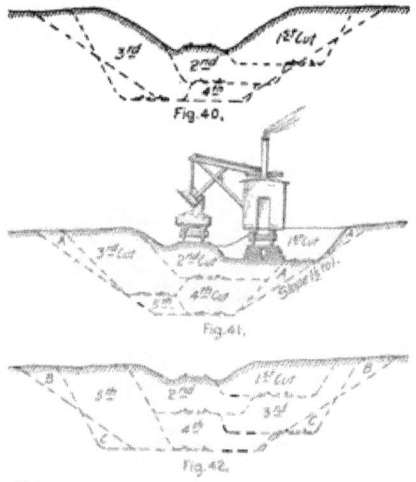

at D and ends at G for the third cut and loads on the pit track in the second cut; the fourth cut is made in a similar way, the machine beginning at E and ending at F, Fig. 36. The fifth and last cut is merely a widening cut, made by loading on the track in the pit of the fourth cut. The material of each cut after the first is loaded on the track laid in the preceding cut. After the completion of the last cut, the permanent subgrade having been reached, the main track is laid on the permanent line, and the small quantity of material obtained from cutting the ditches loaded on cars by hand and taken away for unloading. The most frequent depth of cut made at the summit of grades is about 10 ft. (two cuts), Figs. 38 and 39.

When the main track is on a curve, as frequently happens, an extra cut can often be avoided by slightly changing the alinement of the new main track, and at the same time reducing the degree of curvature, as shown by Figs. 42 1-2 and 43. This

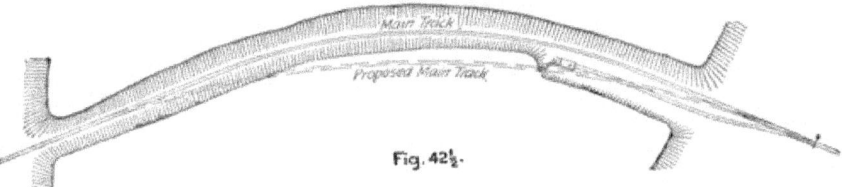

Fig. 42½.

is particularly applicable where an odd number of cuts must be taken to reach the bottom of the new grades. The dipper will cut to a slope of about 1 to 1. When greater slopes are required, it must be done by hand or undercutting resorted to, Sloping by hand is slow and expensive work, impracticably so in all tenacious materials; it has therefore become the exception,

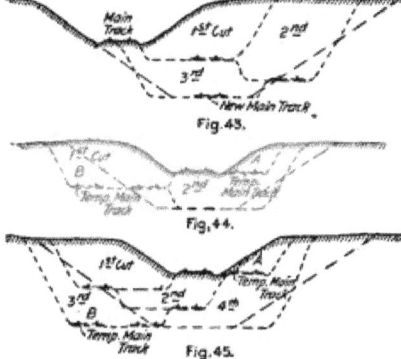

Fig. 43.

Fig. 44.

Fig. 45.

and undercutting the rule. Cuts made in the latter manner sometimes present a rather ragged appearance when just completed, but the irregularities soon merge into a smooth surface as the action of the elements produces the natural slope of the material; the smaller cost amply compensates for the temporary lack of finished appearances. The amount of hand labor necessary where undercutting is not practiced is shown by the sections A in Figs. 38 and 41. This can be entirely avoided by undercutting the slopes, as shown in Figs. 39 and 42; the sections B will slough off within a year or two and most of the material lodge in the spaces C; a small part of this material may roll to the bottom of the cut, and can be removed by loading on cars by hand, or space may be provided for it by making the cut a few feet wider at the bottom. In most cuts for re-

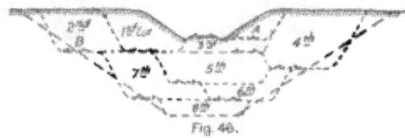

Fig. 46.

ducing grades this extra width must be cut out anyhow to provide room for both steam shovel and loading track.

In reducing grades on railways with a traffic too heavy to permit loading on the main track, a temporary main track must first be built by one of the methods shown in Figs. 23, 24 and 25. The temporary main track, A, Figs. 44, 45 and 46, is then laid, as shown in Fig. 28, to carry the traffic of the road unobstructed. The main track then becomes the loading track for the first cut, and the following cuts are made as shown in

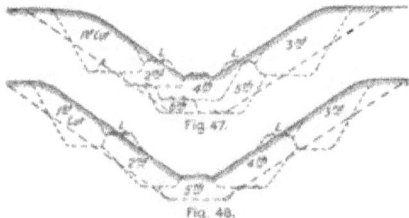

Fig. 47.

Fig. 48.

Figs. 44, 45 and 46. The temporary main track, A, is moved to a second position, B, when the material under it must be cut away. Great care should be taken to arrange the cuts so that the temporary main track will have to be moved as few times as possible, and to attain the lowest level when it is moved. In loose gravel or sandy materials wider bermes and longer slopes must be allowed for the shelf on which the temporary main track rests than are shown in the above figures, but the method of doing the work is essentially the same.

If the depth of the original cut in tenacious materials exceed the height which the dipper can reach, and break down the material above it, the cuts are arranged as shown in Figs. 47, 48 and 49. Temporary loading tracks, L, are built on the side of the slope, and the first cut on each side made by loading on them; the following cuts are then made, as shown on the figures. If the main line traffic is very heavy, it is turned over the temporary main track, A, Fig. 47, until the cut is completed.

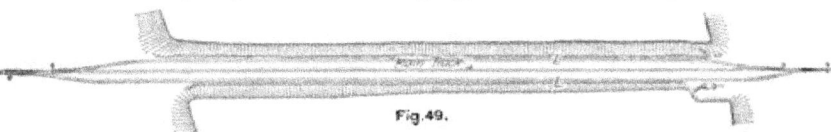

Fig. 49.

The original cuts are not often more than 10 ft. deep, and the section shown in Fig. 45 covers the majority of cases.

On double-track railways the traffic in both directions is generally turned over one track for the length of the new cut, thereby avoiding considerable expense in providing two temporary main tracks.

Each different piece of work presents different conditions; and while the same general principles apply to all, every case

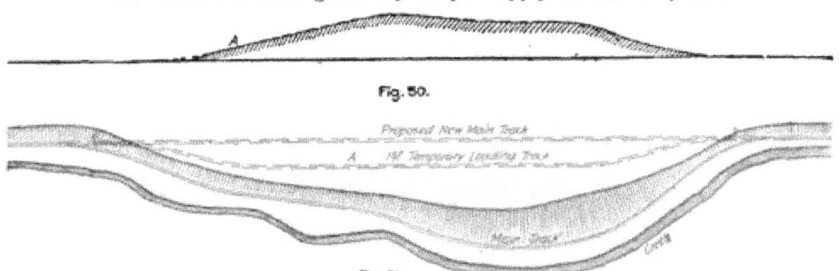

Fig. 50.

Fig. 51.

requires disposition according to its own special circumstances. Great care and study should be exercised in arranging the cuts, to reduce them to the fewest possible number, and avoid shifting, taking up and relaying tracks oftener than absolutely necessary.

Construction Work.—On railways the steam shovel is used mostly in connection with maintenance of way work: loading gravel for ballasting the track, widening cuts, filling trestles, etc., but it is also largely used for various construction work, particularly re-alinements of the main track for reducing grades and curvature. In excavation of this class, thorough cutting should be avoided if possible, for reasons which will be subsequently explained. The work is begun by laying a temporary track, A, Figs. 50, 51 and 52, over the surface of the ground if its natural grade is not too steep to permit operating construction trains over it. Grades up to 6 per cent. (316.8 ft. per mile)

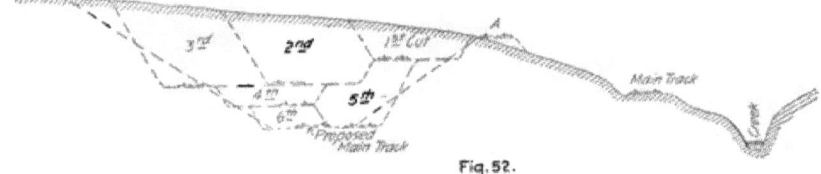

Fig. 52.

can be used. A mogul engine will draw six empty flats over such a grade, a sufficient number of cars to start the work for the short cuts near the summit. The cuts are then made as indicated in Fig. 52.

If the grade of the ground is too steep to operate a track laid on it, one of the three methods may be adopted to obtain a grade for this track:

1. The steam shovel is made to cut a trench between the

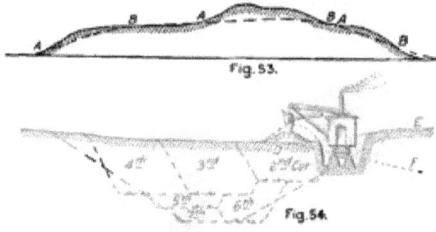

points A and B, Fig. 53, where the slope of the ground is too steep to permit operating a track laid on its surface, and varying in depth from 5 to 10 ft. as may be necessary to attain the desired grade. The excavated material is dumped at D, Fig. 54, to be removed with the next cut. The length of the crane will not permit dumping at E a sufficient distance (20 ft. or more) to obtain a berme and prevent the material washing back into the new cut in the course of time; it must, therefore, be dumped at D and removed as described, unless the slope of the ground is away from the cut, as indicated by the line D F,

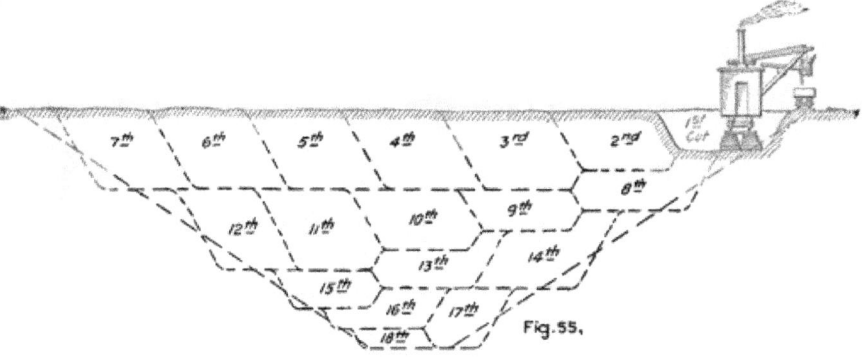

Fig. 54: in such a case the excavated material can be dumped at F.

2. By excavating the trench with teams and scrapers.
3. By through-cutting a trench with the steam shovel, load-

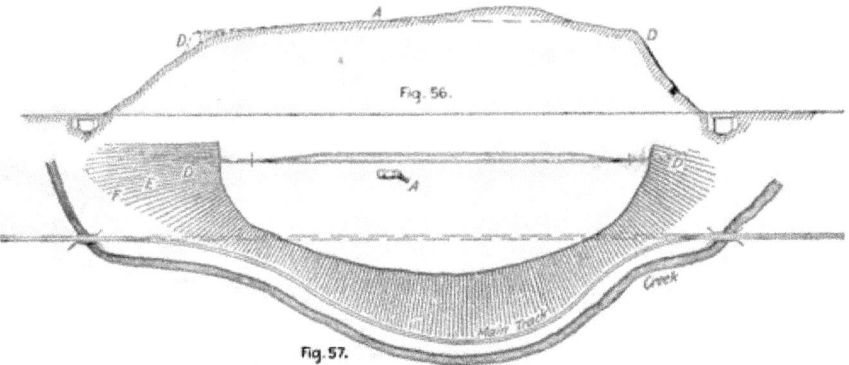

ing the material on small dump cars or wagons, and wasting it at the nearest available place.

After the first loading track has been laid in this trench, the cuts are made as indicated in Fig. 54.

36   STEAM SHOVELS AND STEAM SHOVEL WORK.

When the slope of the ground is too steep to permit a track to be laid on it which can be operated, or to cut a trench for it, as frequently occurs when the excavation passes through a high spur or knoll, Figs. 55, 56 and 57, the steam shovel mounted on standard gage railway tracks cannot be used, and a machine independent of a railway track for transportation must be employed. It is started at A, Figs. 56 and 57, loading small dump cars drawn by horses, and dumping at the nearest

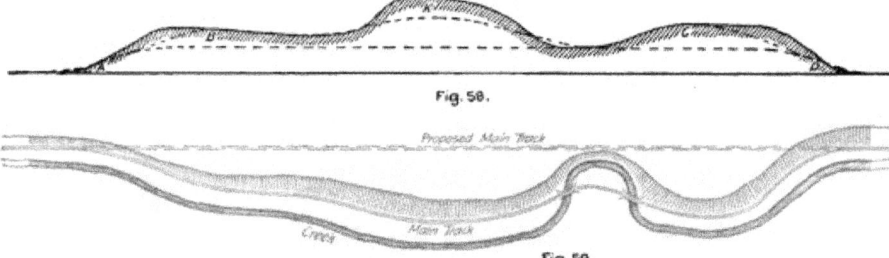

Fig. 58.

Fig. 59.

available place outside of the lines of the new cut, as at D, Figs. 56 and 57. Sometimes wagons are used if the cuts near the top are short and not very deep, so that a temporary standard gage track can soon be run through the cut, and the material loaded on cars. The dumping track at D is changed to E F, etc., Fig. 57, as the machine cuts lower, maintaining a descending grade from the steam shovel.

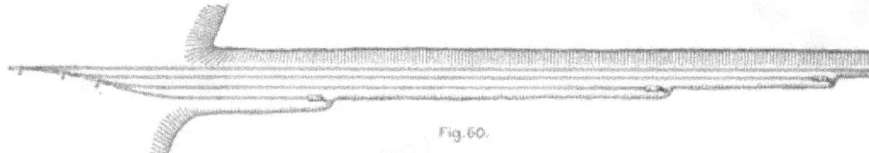

Fig. 60.

In cases of this kind it is often necessary to run the steam shovel up a very steep grade to reach the point where it is to begin work. This can readily be done by attaching one end of a one and a half inch rope to a strong tree and winding the other end around the driving axle. Then starting the running gear the machine can be drawn up grades where it could not otherwise propel itself. As a precautionary measure, it is advisable to use at least two ropes.

A combination of all these methods sometimes becomes necessary, as shown in Figs. 58 and 59. The material in the knoll, K, Fig. 58, is loaded on small dump cars and unloaded at the nearest available place. When this knoll has been cut down sufficiently, and trenches cut between A B and C D, the track A B C D is built, and the excavation proceeded with, as heretofore described. The high points B, K and C are cut down first until the grade of the loading track between B and C is parallel to the grade of the proposed new main track. Cuts nearly 100 ft. in depth and a mile in length have been excavated in this manner. Two and often three steam shovels are employed at the same time, working near the ends of the cut until the through track has been laid, and then following each other, as shown in Fig. 60. As soon as possible, a through track should always be laid, as it greatly increases the capacity for the prompt and efficient handling of the cars.

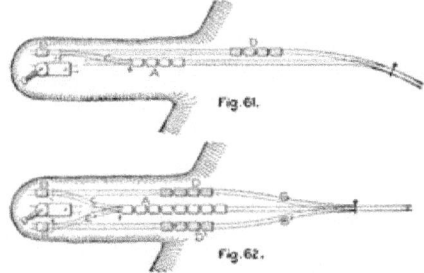

Enough side tracks for storing both empty and loaded cars should be built close to the work, where they can be reached without going out on the main track. Sometimes the pit tracks behind the steam shovels are utilized for this purpose, but these tracks are taken up too often, and should not be depended upon for side tracks, though they may be used as such occasionally.

In through-cutting the material is loaded on small dump cars running on tracks of about 3 ft. gage, drawn by horses, and wasted on some side hill or other nearest available place; this haul seldom exceeds a quarter of a mile in length. In Fig. 61, the empty dump cars standing at A are drawn over the cross-over C by a horse, to be loaded at B; then run to D, and when from four to six cars have been loaded they are taken to the dumping place and unloaded; then returned to A.

In loose materials considerable time is lost in waiting from the time the loaded car is run to D and the next empty brought from A to B. In tenacious materials not nearly so much time is lost, as the dipper cannot be filled so rapidly. This loss of time is largely avoided by arranging double loading tracks, Fig. 62, one on each side of the steam shovel, and connected to a central track for empties by the cross-over C and C' and switches S and S'. Two horses are used, one on each side of the central track, to bring forward the empty cars from A to B, and A to B', and return them to D and D'; these operations are alternately performed, each empty car on one loading track being brought forward while the other is being loaded. The cross-overs C and C' should be kept close to the rear of the steam shovel, and as it advances they must be taken up and relaid; this becomes necessary about once in three days in soft materials and about once a week in hard stuff.

Portable sections of tracks, switches and cross-overs are generally used between the points A and B, and can be relaid very quickly.

Standard gage railway cars cannot be used in thorough cutting, as the track cannot be laid in front of a point at right angles to the post of the steam shovel, and when the track ends there the crane cannot swing back far enough to load the car. Thorough cutting should be avoided if possible, the cost due to the loss of time in switching cars, relaying tracks, extra horses and men, etc., makes it more expensive than excavating from a side cut.

In excavating canals, harbor and dockwork, stripping coal-fields, stone quarries, grading for new city additions, and other work not connected with a railway, as well as railway construction and re-alinement work which is inaccessible to a railway track in its early stages, the general manner of using the steam shovel is the same as for railway work; varying only in details, depending upon the means of disposing of the loaded material, by wagons, carts or dump cars, and the use or waste of this material.

Although the steam shovel is employed mostly on railway work, it is not exclusively a railway machine. It is already largely used on other work, and its use in this direction is rapidly extending, especially on the increasing number of extensive public works in the vicinity of large cities.

The most economical height of cut varies greatly with the nature of the material. In dry clay, loam and other dry materials which can be broken down readily with a bar or iron pointed pole (Fig. 17), cuts of 25 to 30 ft. in height are usually taken. In harder and more tenacious materials it should not exceed the height to which the dipper can be raised, 14 to 20 ft., varying with the size of the machine. In sand and loose gravel which easily falls down to the machine heights up to 60 ft. are common, and side-hill cuts in loose gravel up to 300 ft. in height have been taken. In such cases, and also in the removal of landslides, great care must be taken to avoid an avalanche of the material burying the machine when the toe of the slope is cut away. The pit track should always be kept close up to the sections of track under the steam shovel, so that it can be quickly withdrawn when necessary. As a general rule, the higher the cut the better, as the machine can then load the greatest amount of material between each advance, and lose the least possible amount of time. Each forward move of the machine requires from three to ten minutes, depending upon the height of blocking, if any, it is working on; this is a dead loss, as no cars or wagons can be loaded during that interval.

Powder and dynamite are frequently used to good advantage to shatter the harder materials before excavating. When thus broken up about twice the amount of these materials can be loaded in a day. Great care must be exercised in the quantity of the explosive used, and in the location of the drill holes to prevent injury to the steam shovel. The explosives should be stored in a safe place, preferably in a vault at some distance from the place where they are to be used.

The use of dynamite is confined mostly to bowlders, ledges of rock and stumps of trees, while powder is generally used for hardpan, shale, slate, cemented gravel and hard clays. For the latter materials dynamite is usually too powerful, as instead of merely lifting and loosening them, as desired, it shatters shale and slate into fragments, and compresses the other materials about it, forming a "cistern" from 3 to 5 ft. in diameter, as shown in Fig. 63. Sometimes small quantities of it are used specially for this purpose to make room for a large charge of powder at the bottom of the drill hole, where its explosion will have the most effect in loosening the superincumbent material. A charge of one-quarter to one-half of an ordinary dynamite cartridge will usually blow out a "cistern" large enough to contain from one-half to one keg of powder, Fig. 64.

The depths of the drill holes in these materials vary from 4 to 20 ft.; they are made with a drill, or, in the softer materials, with an auger similar to a plank auger, generally about 2 ins. diameter, with extension pieces for deep holes, as shown in Fig. 65. Crowbars and wooden and iron wedges are also often used in breaking down overhanging material when it cannot quite be reached by the dipper.

The excavation of materials for which powder or dynamite are used to loosen them requires a powerful machine, with a strongly built, medium size dipper. A small or lightly built machine giving good satisfaction in soft materials would prove an utter failure here.

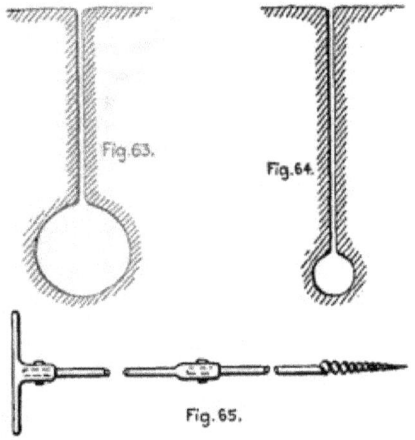

Fig. 63. Fig. 64. Fig. 65.

Assuming good management and a competent crew, the daily output of a steam shovel depends mostly upon the nature of the material excavated; it is also somewhat dependent upon the height and width of the face of the cutting, and largely upon the facilities for disposing of the loaded material, and keeping the machine almost constantly at work by an ample supply of empty cars and wagons. Although these varying conditions differ on each piece of work, the probable output of a machine for a given excavation can be closely estimated by good judgment based on previous experience with similar work. The average daily output in different kinds of materials, and under average, favorable and unfavorable, conditions, as described above, is shown in Table II.:

TABLE II.

Capacity of dipper.	Delay, hours.*	Sand. Cu. yds.	Loose gravel. Cu. yds.	Dry loam. Cu. yds.	Dry clay. Cu. yds.	Damp clay. Cu. yds.
2½ cu. yds	1 Good	2,400	2,400	2,000	1,800	1,200
"	5 Poor	1,200	1,200	1,000	900	600
"	2½ Avg	1,800	1,800	1,500	1,350	900
1¾ cu. yds	1 Good	1,600	1,600	1,200	1,000	800
"	5 Poor	800	800	600	500	400
"	2½ Avg	1,200	1,200	900	750	600
1 cu. yd	1 Good	1,000	1,000	800	700	500
"	5 Poor	500	500	400	350	250
"	2½ Avg	750	750	600	525	375

* The delay in hours is the time lost in moving forward and waiting for empty cars.

TABLE II.—Continued.

Capacity of dipper.	Delay, hours.*	Stiff blue clay. Cu. yds.	Hard pan. Cu. yds.	Loosened by explosives.		
				Mixed clay and boulders. Cu. yds.	Loose rock. Cu. yds.	Cemented gravel. Cu. yds.
2½ cu. yds	1 Good	800	600	600	600	600
"	5 Poor	400	300	300	300	300
"	2½ Avg	600	450	450	450	450
1¾ cu. yds	1 Good	600	400	400	400	400
"	5 Poor	300	200	200	200	200
"	2½ Avg	450	300	300	300	300
1 cu. yd	1 Good	450	300	300	300	300
"	5 Poor	200	150	150	150	150
"	2½ Avg	300	225	225	225	225

* The delay in hours is the time lost in moving forward and waiting for empty cars.

Part III.—Disposition of Material.

Loading the Material for Transportation.—The material excavated by a steam shovel is loaded on cars, wagons or carts. On railway work it is usually loaded on dump or flat cars. On other construction work small dump cars are most generally used, and sometimes wagons or carts.

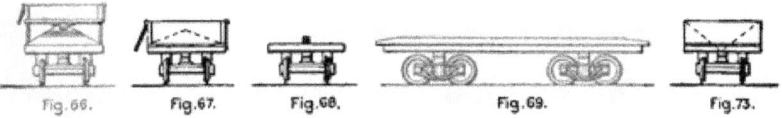

Fig. 66.   Fig. 67.   Fig. 68.   Fig. 69.   Fig. 73.

Standard gage railway dump cars, Figs. 66 and 67, have nearly gone out of use. They were replaced by the center ridge flat car, Figs. 68 and 69, and it in turn has been replaced by the ordinary flat car. Dump cars are of two styles, dumping either by tipping, Fig. 66, or by means of a hinged sideboard opening on an inclined floor, Fig. 67. Both are heavy, clumsy, costly and can be used for scarcely any other purpose, often standing idle from six to eight months of the year. They dump dry materials very rapidly, but are often slow in discharging damp, tenacious materials, especially in the hinged sideboard car, whose floor slope is often not sufficient to permit the material to slip out quickly, and the material must then be pushed out, thus causing much delay. The greatest objection to these cars is that they can be used for scarcely any other purpose, on most railways for no other purpose; and there is not sufficient work for them to justify keeping the necessary number on hand for the ordinary work in this line. They were replaced by the center ridge car, Figs. 68 and 69, as above noted, which is merely an ordinary flat car with a timber 4 by 6 ins. bolted on its floor along the center line, serving as a guide for a plow, Fig. 70, drawn over it by the locomotive, thereby unloading the material. The ridge timber is slightly pointed at both ends to assist in guiding the plow onto the car as it passes from one

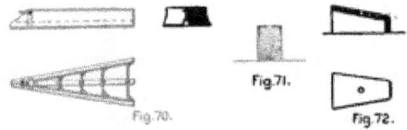

car to another. The top edges of the ridge are sometimes protected by angle irons, as in Fig. 71, and the points by cast iron caps, Fig. 72. By taking off the center ridge this car can readily be restored to general service after completing the steam shovel work. The center dump car, shown in Fig. 73, is used only for gravel ballasting where the material is wanted delivered between the rails.

The brakes are placed on one side of the car, as shown in Figs. 74 and 75. When boulders, loose rock, etc., are to be unloaded, the brake staff is set in a socket, Fig. 76, and taken out before the plow is started. This avoids bending or breaking the staff in case any stone should be wedged between it and the moving plow. Sometimes the socket is used with the brake at its ordinary place at the end of the car; in such a case it must always be taken out before the plow reaches it.

The plow, Fig. 70, is built of heavy plate and angle iron, strongly braced, and headed by a cast steel point, to which the cable is attached. The sides are curved outward at the bottom, working under the material and pushing it aside as the plow is drawn along, and held down on the car by the weight of the material and the partly downward pull of the cable at its point. Short pieces of old rails and other scrap iron are also often placed on the plow to help hold it down on the car when very tenacious materials are to be unloaded. The groove extending along the center line on the bottom fits over the ridge timber on the car, and forms the guide by which its movement is directed. Small stones, protruding bolts, slivered ridge timbers and other obstructions in the groove of the plow sometimes wedge the point fast, and before the engine can be stopped, the plow is turned up on its point, and falling to either side, tumbles off the car. The weight and elasticity of the cable is often sufficient to draw the plow half a car-length after the engine has been stopped, and it is often difficult to stop the plow quick

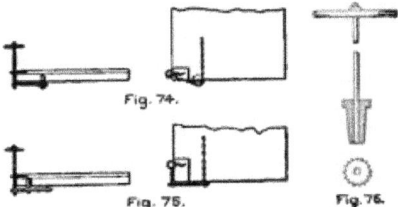

enough to prevent upsetting when obstructions occur, although the speed is usually only two to three miles per hour. The unloading nearly always occurs on trestles or embankments, and when the plow is thrown off the car, its replacement often requires much time and labor, sometimes even making the services of the wrecking car necessary. This difficulty is very likely to occur when unloading on curves, where one side of the point of the groove presses against the ridge timber. This plow unloads the material equally on both sides of the car, as it is wanted in filling trestles, raising embankments, tracks, etc.; but it cannot be used to advantage where the material is wanted on one side only, as in widening embankments for double track, side tracks, yards, station grounds, etc.

The many objections to the center ridge car are almost entirely avoided by the use of the Barnhart plow, Fig. 77, employing the ordinary flat car without any preparations except changing the brake staffs to one side or placing them in sockets at their ordinary places and inserting short stakes in the stake pockets, permitting the immediate use of the car for general service if necessity should so require. This plow is also built of heavy plate and angle irons, strongly braced, and headed by a cast steel point to which the cable is attached; it is preceded and followed by guiding sleds attached to it by adjustable hinges and guided over the car by the stakes in the stake pockets, which are indicated by the dotted lines. The usual speed at

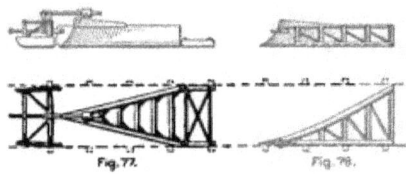

which it is drawn over the car is about four miles per hour, but in loose gravel it can safely be drawn at a speed of six miles per hour. On straight track it is scarcely ever thrown off the car unless carelessly handled, and it works equally well on curves when the usual means are adopted to maintain a tangential pull of the cable, as will be subsequently described. Two styles of the Barnhart plow are in use: One unloading on both sides of the car, and called the center plow, Fig. 77; and the other unloading on one side only and called the side plow, Fig. 78.

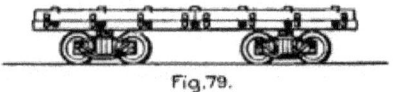

On all but very small pieces of work the cars should be provided with hinged drop sideboards, Fig. 79, using either of the arrangements shown in Figs. 80 and 81, which will enable them to carry 12 to 14 cu. yds. instead of 6 or 7. The side boards are made in two pieces on each side of the car, Fig. 79. Those shown in Fig. 80 are used for both center and side plows; they can be quickly dropped by a man walking along the train, after arriving at the unloading place and striking the hook A an upward blow with a light hammer. The boards are hooked up again after the cars have been returned to the steam shovel pit. The side boards shown in Fig. 81 are used where the side plow only is used. Here the board on one side only (the unloading side) is hinged (or chained), and dropped by pulling out the pin B, thus leaving that side of the car entirely unobstructed for unloading the material; the board on the other side of the car is bolted to the stake pocket and is not moved.

The cars should also be provided with sheet iron aprons, Figs. 82 and 83, extending from the end of one car onto the floor of the next, to prevent the material from falling on the track between the cars as the plow is drawn over them, and delaying the departure of the train until it can be shoveled out. These aprons are made either in two pieces, Fig. 82, or in one piece only, Fig. 83. The former are more easily handled, and

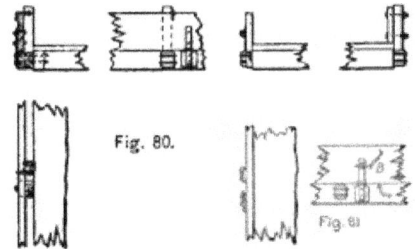

Fig. 80.   Fig. 81.

permit access to the coupling of the cars without lifting the apron. Very little material drops on the track when the aprons and the center plow are used. The single apron is used mostly in connection with the side plow.

The number of cars and engines required for each steam shovel to keep it in nearly constant operation depends upon the nature of the material excavated, the length of haul, and the density of other traffic upon the main line. This number must be determined by accompanying circumstances in each case; ordinarily, however, it averages about as given in Table III.:

TABLE III.

	In the steam shovel pit.		On the road up to—					
			10 miles.		25 miles.		50 miles.		75 miles.
	Loco.	Cars.	Loco.	Cars.	Loco.	Cars.	Loco.	Cars.	Loco.	Cars.
Loose gravel	1	30	1	30	2	60	3	90	4	12
Dry clay	1	22	1	22	2	49	..	..	..	..
Damp stiff clay	1	18	1	18	2	36	..	..	..	..
Hardpan, cemented gravel, etc., loosened by explosives	1	16	1	16	2	32	..	..	..	..

The length of haul usually ranges from 2 to 15 miles; it seldom exceeds 25 miles for any material except gravel ballast, where hauls of 75 miles are frequent, and sometimes reach 200 miles.

On hauls exceeding 25 miles the full number of cars and engines required can seldom be obtained, and the output of the

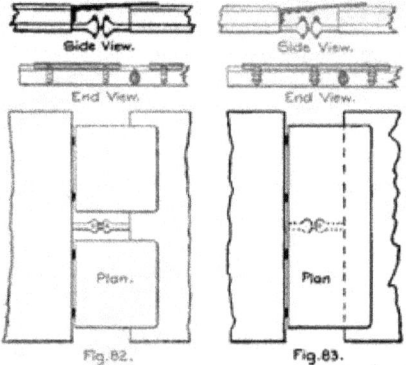

steam shovel is correspondingly decreased. The delay in returning empty cars due to detentions from other trains is the great trouble most keenly felt in steam shovel work on railways in operation. The so-called "mud train" is generally considered an outcast, and is usually the last train to receive the dis-

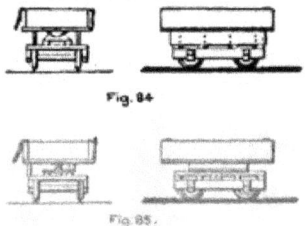

patcher's attention for an order to the road. These delays are daily occurrences, and it is quite an exceptional case when the machine is amply supplied with empty cars. The record of most steam shovels on such work is therefore a rather poor one, when the machine really made a good showing for the crippled condition of its car service. Some of these delays can be avoided or shortened by stationing a telegraph operator at the outgoing end of the pit, and on all but very small pieces of work his wages will be many times balanced by the time gained in keeping the whole plant moving, by obtaining train orders quicker, and remaining constantly informed of the whereabouts of the construction and other trains, and regulating the work in the pit accordingly.

For general construction work where the excavated material is not loaded on standard gage railway cars, small dump cars, Figs. 84 and 85, are generally used. They are more economical than wagons or carts, which are employed only in special cases, mostly in cities, where the material must be hauled some distance over several intersecting streets, and where a track will not be allowed; or for very small jobs with a long haul which would not justify building a track.

The gage of these tracks is usually 2 1-2 or 3 ft., sometimes 2 ft. or even 1 1-2 ft. only; the latter gages are not often used, and the 3-ft. gage is usually preferred.

The rails most generally used weigh 20 lbs. per yd. Although these tracks are only temporary their construction should be fairly substantial; but they are often built in an exceedingly careless and insecure manner, causing a great waste of power in pulling the cars over them, and resulting in frequent delays, due to derailments. The grade is usually arranged so that the loaded cars will run downhill by gravity, and only the empty cars need be drawn back to the pit. On small work, horses or mules are used to pull the cars, but on large jobs small locomotives are employed. Small dump cars vary in capacity from 1 to 3 cu. yds., the latter size being most generally used. The side dump car, Fig. 84, dumps on either side. The rotary dump car, Fig. 85, unloads on either side or end; the box can be turned around horizontally, revolving about a vertical pin in a turntable on the frame; they are used mostly in dumping off the end of a fill.

In making fills it is nearly always the best plan to build a temporary trestle of round pieces of beech, cottonwood or other cheap trees, old bridge or building timber, or other second-class lumber, and then filling in with the side dump cars. By adopting this plan the unloading will progress much more rapidly than by dumping from the end of a fill, where only one car at a time can be unloaded. These trestles are inexpensive, and the saving in labor and time in making the fill will amply repay their cost.

Unloading the Material.—On railways the unloading is seldom done by slow and expensive hand labor with the shovel; sometimes dump cars are employed, but in most cases flat cars and the plow are used. The trains consist of 10 to 30 cars. The car carrying the plow is attached to the rear of the train at the nearest side track to the unloading place, if it is not over 10 miles from the steam shovel pit this car is generally carried back and forth to avoid an extra stop to couple it on the train at the side track. One end of a steel wire cable is then hooked to the plow and the other end (which is attached to an ordinary car coupling link) coupled to a car or the engine. Usually this cable is about 400 ft. long and extends over 12 cars. The

Fig. 86.

Fig. 87

Fig. 88.

Fig. 89.

brakes on these cars are then set up tight and the engine started with the forward cars, Fig. 86. In very tenacious or partially frozen material the rear cars are sometimes pulled along by the plow; the wheels are then blocked with pieces of wood or with stones; sometimes it is even necessary to chain a few of these cars to the track to prevent the rear lot of cars from moving. After the plow has been started, it is drawn along slowly until it arrives on the last car, Fig. 87. The engine is then stopped and backed up a few feet to permit the cable to be thrown on one side of the track, Fig. 88. The train is then backed up again and coupled to the unloaded cars, when four to six men throw the cable on the next loaded cars, Fig. 89, coupling its forward end to a car or to the engine if the cable is long enough. The operation is then repeated until all but the car next to the engine is unloaded; this car carries the plow and is the first car to be unloaded by the next train. The ends of the cable are then detached from engine and plow, thrown to one side of the track, and left there for the next train to pick up and use in the same manner.

When filling a trestle the cable cannot be thrown on one side, as described, but must be unhooked from the plow (the rear lot of cars being left standing on the trestle), dragged across the trestle, and there thrown to one side. The forward lot of cars is then backed up until its rear car is opposite the rear end of the cable, when it is loaded, the train backed up, coupled and unloaded, as before described. After unloading the train the cable must again be dragged beyond the trestle, and there thrown to one side of the track and left for the next train. The time required for unloading varies from 10 to 30 minutes, de-

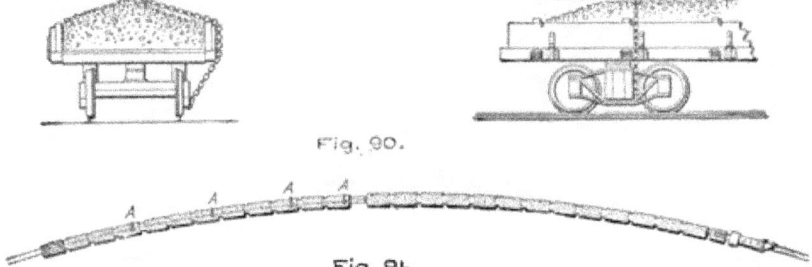

Fig. 90.

Fig. 91.

pending upon the nature of the material and the number of cars, and averages about 20 minutes, doing as much work in that time as 20 men can do in a day.

When unloading on curves the operations are delayed by the necessity of using snatch blocks on the cars to insure a nearly tangential pull of the cable and avoid pulling the plow off the car. These blocks are applied as shown in Fig. 90, and at A, Fig. 91. They are hooked to long chains extending over the car and fastened to the bolster or arch bar of the truck. The number of snatch blocks required depends upon the degree of the curve and the length of the cable; generally four to six blocks, one to every third car, are enough. As the plow approaches one of these blocks it must be stopped, block and chain removed and transferred forward for use at that end of the train. The other operations of unloading are the same as when on straight track. The time required in unloading on curves varies from 20 minutes to an hour, and averages about 40 minutes, doing as much work in that time as 20 men can do in a day.

The steel wire cables used vary from 1 in. to 1 1-2 ins. diameter. The former are used for unloading loose gravel and sandy material; they are light and easily handled, but cannot bear much jerking. The most usual size is 1 1-4 ins. diameter. Heavier cables require too many men (six to eight) to load them on the car preparatory to starting the plow.

One of the heaviest locomotives on the road (preferably one of the consolidation type) should be used for drawing the plow over the cars. These engines are generally able to keep the plow moving with a strong steady pull, avoiding the necessity of taking a run to start the plow, and all injurious jerking of the cable, which frequently breaks it. For tenacious materials and where the haul is not more than 25 miles, it is often good policy to keep one heavy engine at this work, the other engines merely hauling the trains; this can generally be arranged so that no more engines are used than if each engine were to unload its own train. Sometimes two light engines are used for this purpose, but they can seldom move in perfect unison and more or less jerking is the result. Unfortunately the engines for the "mud trains" are not always in the best working order; they are mostly those which are about to go into the shops for turning down the tires or for general repairs, and are not in fit condition for general traffic, but still considered good enough for this service. Expensive delays due to badly working engines are frequently the result.

The locomotive in the steam shovel pit should always be equipped with a steam or air driver brake to assist in quickly stopping the cars at exactly the right place when setting them for loading by the steam shovel. For the same reason the brakeman should be allowed to use short sticks in the brake-wheels to obtain a greater leverage in turning them.

Both engine and train crews should be changed as little as possible and they should retain their respective trains in the pit on the road or at the dump. Most of the men dislike the "mud train" service, but some (especially the older ones) are glad to get a steady job with a full night's rest, and these are the men to be chosen. They take an interest in the success of the work, and soon acquire an expertness in handling cars, plows, etc., that makes them worth twice as much as the inexperienced or unwilling ones. The wages of these men should be equalized to average the same as the men on the road in other service, otherwise dissatisfaction and indifference are sure to result.

The machine shown in Fig. 92 has lately come into use for pulling the plow over the cars to unload them. This is merely a double cylinder (10 by 12 ins.) reversible hoisting engine, resting on a heavy cast iron bedplate attached to the floor of a box car. Steam is supplied to the engine from the locomotive of the train, which is coupled to this car when the unloading is to begin. With this machine there is no injurious jerking of the cable, and consequently very little breakages or delays, and heavy loads of 15 cu. yds. of tenacious material are readily plowed off the cars in a more satisfactory manner than can be done by any one or two locomotives. Blocking the wheels or

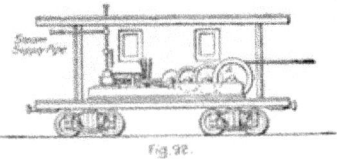

Fig. 92.

chaining cars to the track need not be resorted to; the cars cannot move, for the machine pulls the plow toward itself and the strain is resisted by the cars between them. If it is desired to scatter small quantities of material along the track, as it is often wanted in surfacing or raising track, both plow and train are moved in the same direction at the same or varying speeds, as may be necessary to unload the required amount of material. If a large quantity of material is wanted within a short distance, as usually happens on washouts, train and plow are moved in opposite directions. By moving them in this manner at the same speed, a whole train can be unloaded at any desired spot. Where two locomotives must be used to pull the plow over the cars, the use of this machine will dispense with one of them, and do the work in half the time. On large jobs it should not be missing. The cable is wound around the drum, A, Fig. 92, and must be long enough to extend over the whole length of

52   STEAM SHOVELS AND STEAM SHOVEL WORK.

the train. A steel wire cable 1 1-8 ins. diameter is generally used; but for loose gravel a 1-in. cable is amply strong enough

The steam shovel can be operated continuously throughout the year in all kinds of weather, though operations are often suspended in extremely cold weather. When working in cold weather the face of the bank sometimes **freezes** during the **night to the depth of** 3 to 6 ins., **but** this crust is easily broken in the morning by a few small charges of powder, and then **the material can be excavated as** easily **as at** any other season.

In freezing weather the floors of the cars should be sprinkled with brine just before loading; the brine is kept in barrels at the head of the machine, and one man using an ordinary garden

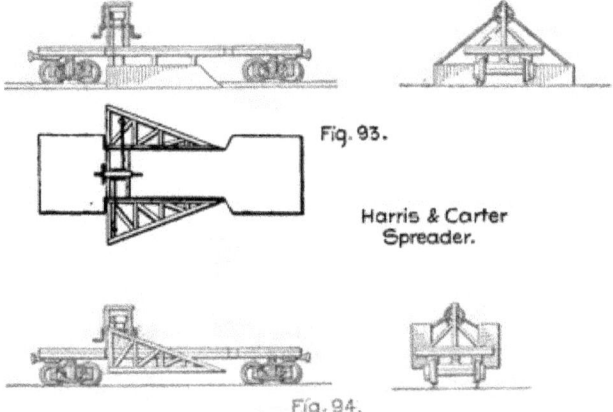

Harris & Carter Spreader.

Fig. 93.

Fig. 94.

sprinkling can is detailed for the work. This prevents the material from freezing to the floor of the car for three to four hours, and allows it to slip off readily when the plow is put in operation. No train should be left standing over night without unloading. The brine will not prevent freezing for this length of time, and to unload **one car of the frozen** stuff **requires a day's labor of** four to six men.

Distributing the Material **After** Unloading.—In widening embankments for side tracks, **double track,** yard **and** station grounds, etc., the material is unloaded, as described above, forming **a ridge on both** sides **of the** track if unloaded with the **center plow, or on one side of the track** only if unloaded with the

STEAM SHOVELS AND STEAM SHOVEL WORK.   53

side plow. This material is sometimes leveled off by hand, a very slow and expensive job, but generally it is done with a leveler or spreader, Figs. 93 to 96.

In the Harris & Carter spreader, Figs. 93 and 94, the car body is cut away between the trucks to receive the two wings which level or spread the material. One or both wings can be used, and they can be raised and lowered to adjust them to any height of new embankment wanted. They will spread the material for a distance of 3 ft. from the rail. When shipping the spreader over the road the wings are drawn up by a hand wind-

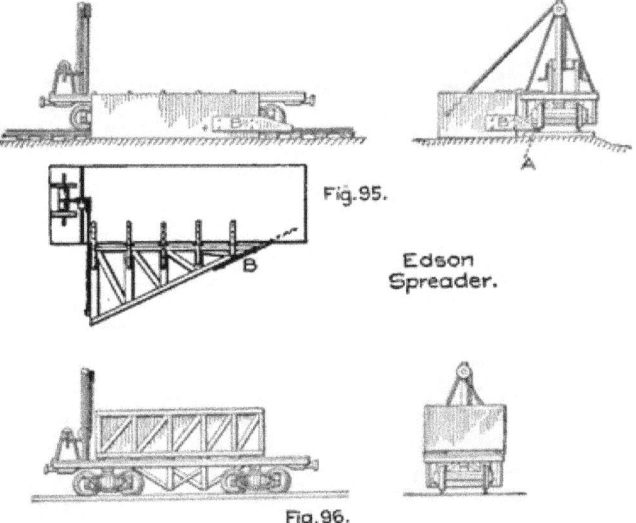

Edson Spreader.

Fig. 96.

lass, revolving about hinges fixed to the braces under the floor of the car, as shown in Fig. 94. In this position the clearance is the same as that of an ordinary passenger car.

The Edson spreader, Figs. 95 and 96, has only one wing, attached to an ordinary flat car, and arranged to raise and lower to adjust it to any height of new embankment wanted. The wheel, A, bears against the head of the rail, forming a brace where one is most needed, and greatly assists in preventing a derailment when hard or tenacious materials are suddenly encountered. The wing, braces, windlass, etc., are so constructed that they can be readily removed from the car, thereby restoring it to general service on completion of the work in hand. This spreader is used mostly in connection with the side plow; it will level the material for a distance of 15 ft. from the rail, wide enough to permit laying a side track from which the embankment can be further widened. Only one side at a time can be widened with this spreader. If it is desired to widen the embankment on both sides of the track, one side is completed first; the cars and spreader are then turned around on the nearest turntable or Y-track, and the other side widened by drawing the spreader in the opposite direction. If the cars are not provided with aprons they need not be turned around. This spreader is generally arranged to cut about 6 ins. below the bottom of the ties of the main track, thereby forming the subgrade for the side track, and maintaining proper drainage of the main track. The apron, B, is bolted on the spreader, and serves to remove any loose material which may fall on the track between the rail and the ends of the ties. When shipping the spreader over the road, Fig. 96, it is drawn up by a hand windlass revolving about hinges on the side sill of the car and folded down on it; in this position it will clear anything that other cars can pass.

The cars of both styles of spreaders are loaded with old rails, frogs, scrap iron, etc., to hold them down and prevent derailments when hard or tenacious materials are suddenly encountered. Loads of five to ten tons are generally sufficient, though loads up to 15 tons are sometimes required.

Spreaders are usually drawn at a speed of six to eight miles per hour; in loose gravel the speed often reaches 10 miles per hour. They will level off a ridge a mile in length in six to ten minutes, doing as much work in that time as 100 men can do in a day.

The spreader is usually stationed in the nearest side track to the unloading place. Frequently it can be hauled between this track and the dump without raising it, or raising it only partially to clear depot platforms, switch stands and other obstructions and thereby avoid the necessity of folding it down on the car while passing between these points.

Ordinarily the spreading is done by the last train before the close of the day. In cold weather or on short dumps it must be done oftener; either to prevent freezing, or to make room for the unloading material which would otherwise pile up too high for easy spreading, or be liable to roll back on the track and obstruct it for the next train. In using the spreader it is coupled to the rear of the car carrying the plow, and after the train has been unloaded it is pulled over the length of the ridge of material unloaded from its own and preceding trains, as shown in Figs. 97 and 98.

Part IV.—Cost of Steam Shovel Work.

The cost of steam shovel work varies greatly with the different conditions affecting each piece of work. It depends mainly upon the nature of the material, its location, the capacity and efficiency of the steam shovel, and the supply of empty cars or wagons. The efficiency of a steam shovel is not necessarily proportional to its capacity, but to the amount of work done compared to its cost; and while the amount of work done is generally larger in the machines of larger capacities, this advantage may be more than balanced by the greater cost of operation, including the cost of labor, fuel, supplies and repairs, etc. Machines of the largest capacity, with dipper of 2 1-2 cu. yds. capacity, are employed mostly in excavating soft materials, especially in loading gravel for ballasting. Machines of medium capacity are usually the most efficient for general construction work.

The average daily operating expenses of a steam shovel of medium capacity are about as follows:

One engineman	$4.00	
One cranesman	3.50	
One fireman	2.00	
Four pitmen at $1.50	6.00	
Wages of crew	$15.50	$15.50
One ton coal	$3.00	
Oil and waste	.75	
Water	.50	
Fuel and supplies	$4.25	$19.75
Interest on capital, $6,000, at 6%	$1.00	
Depreciation at 10%	2.00	
Repairs	1.00	
	$4.00	
Total daily expense with regular crew		$23.75

This will suffice for loading loose gravel; in the harder materials ordinarily occurring on construction work the following daily expenses must be met:

56 STEAM SHOVELS AND STEAM SHOVEL WORK.

Expenses of regular crew		$23.75
Foreman	$5.00	
Two pole (or bank) men at $1.50	3.00	
Two extra men at $1.50	3.00	
One night watchman	1.50	
Powder and dynamite	1.00	
	$13.50	
Daily expenses on average construction work		$37.25

To the above must be added the expense of transporting the machine to the work, and returning.

The cost of hauling is also a variable item; it depends mostly upon the length of the haul, and on railways very largely upon the delays met with in going to and from the dumping place. On construction work it is seldom less than 3 cts. per cu. yd., and sometimes reaches 10 cts. On railways it is not often below 4 cts. for hauls up to 10 miles in length, and may reach 50 cts. or more for hauls of 75 miles or farther.

Dumping is a very small item where small dump cars are used on construction work, and does not exceed 1-2 ct. per cu. yd.

Fig. 97.

Fig. 98.

When wagons are used it will average about 1 1-2 cts. On railways the cost of unloading with the plow varies somewhat, depending upon the kind of material; it averages about 1-2 ct. per cu. yd. Unloading by hand averages 6 cts.

On railway work, where the spreader is used, the average cost of leveling the material for widening embankments is only 0.1 ct. per cu. yd.; spreading it by hand will range from 5 to 20 cts. per cu. yd. for widths of 5 to 15 ft. from the unloading track.

The total cost per cu. yd. of excavating and loading, hauling and dumping different kinds of materials with the most usual length of haul averages about as follows:

	Loading. Cents.	Hauling. Cents.	Dumping. Cent.	Total. Cents.
Sand and loose gravel	3	1 to 10	½	7½ to 13½
Loam	3½	"	"	8 to 14
Dry clay	4	"	"	8½ to 14½
Damp clay	6	"	"	10½ to 16½
Stiff blue clay	8	"	"	12½ to 18½
Cemented gravel, hardpan, etc., materials loosened by explosives	10 to 16	"	"	14½ to 26½

The steam shovel will do the work of 60 to 120 men, saving from 5 to 25 cts. per cu. yd. of material excavated and loaded. The gain is proportionally much greater in the harder, and particularly in the more tenacious materials. The machine is not adapted to small jobs, and is seldom worked in cuts of less than 8 ft. in depth; nor is it cheaper than hand and team labor on such small jobs, but on nearly all large work it is much cheaper and faster; and last, though not least, its use largely reduces the number of laborers required, and hence the probability of strikes and other labor troubles.

APPENDIX.

ACTUAL COST OF STEAM SHOVEL WORK.

(From an article in Engineering News, June 9, 1888, we take the following particulars of reports on the actual cost of steam shovel work, and these reports show how variable is the cost of excavating, depending, as it does, upon delay, unavoidable on every line of railway, upon the weather, character of the material, length of haul, and many other conditions. When conditions are favorable as to material, prompt and short hauling, with no delays, the results show a very large increase in the output, and often a decrease in cost.—Ed. Eng. News.)

From a report of the General Roadmaster of the New York Central & Hudson River R. R. of work done by two shovels on the Eastern and Western divisions, we find the largest day's work for one shovel at Yost's pit was 174 cars, the average for the month of August being 121 cars per day and for July 116 cars per day. It could have made a larger average than this with twenty more cars, as the trains making long runs could not keep cars in the pit. The largest day's work at Bergen pit with one machine was 156 carloads, the June average being 117 cars and the July 116 cars per day, and for two weeks in August 134 cars per day. At this pit they came in contact with cement, hard-pan, and very coarse material. At Yost's pit they have loaded 10,511 cars in four months up to Aug. 1. Figuring these at 9 yds. per car, which is low, makes 94,599 yds. The cost of delivering on roadbed was $5,261.25, or about 5½ cts. per yd. The average cost for handling by men loading and unloading is 14 cts. per yd.

The report on a machine working in New Mexico on the Atchison, Topeka & Santa Fe R. R. says: "In cemented gravel, we find no difficulty, under favorable circumstances, in loading 75 to 600 cars per day, at a cost not to exceed 10 cts. per cu. yd."

The engineer of the Cleveland, Mt. Vernon & Delaware R. R. gives some statements as to the cost and amount of some excavating work done under his direction. This shovel worked about 5½ months in stiff clay, as follows:

March loaded 1154 cars, worked 24 days.				Sept. loaded 1556 cars, worked 23 days.				
July	"	955	"	"	24	"	Oct.	"	1552	"	"	27	"
Aug.	"	1157	"	"	22	"	Nov.	"	539	"	"	12	"

Total, 6,915 cars, 41,490 cu. yds. Greatest number of cars loaded in a single day, 97. Shovel supposed to work ten hours a day, but did not average more than 6½ hours on account of waiting for cars. Carloads average 6 cu. yds. per car. Average cost of loading, 3 cts. per cu. yd., including expense of all men, shovel, oil, waste, etc. Loaded, hauled material, and unloaded at a distance of ten miles from pit, at 10 cts. per yd., including all costs, shovel, use of cars, engines and crews. A 20-mile haul on this road cost 15 cts. per yd., and a 30-mile haul about 20 cts. per yd., while on some roads a 30-mile haul costs over 75 cts. per yd., depending on the frequency of trains.

The following report from the superintendent of the Sioux City & Pacific Ry. gives the operations of a shovel for nine months working in a yellow clay bank from 30 to 40 ft. in length, and with a one-mile haul: " The total number of cars loaded was 31,420 in 209 days, giving an average of 150¼ cars per day. The greatest number of cars loaded in one day was 275, with an average of 6 cu. yds. per car. The average cost of loading per cu. yd. is 6½ cts., including expense of all men about shovel, and shifting of shovel track. Average cost of unloading with one-mile haul, 7.8 cts., including wages of all men with trains and engines, use of cars and locomotives, with all supplies and repairs of same, making a total cost of 14.3 cts. per cu. yd. or 85.8 cts. per car delivered on track."

A report showing the largest amount of work, with the most complete detail as to the expense of operation was furnished by the resident engineer of the Missouri Valley & Blair Railway & Bridge Co., contractors for the Chicago & Northwestern Ry. bridge across the Missouri River at Missouri Valley, Ia., the material excavated being used in the approaches to the bridge. The work, a tabulated statement of which is given in Table IV., was done under the most favorable circumstances, with but few delays, and with but one locomotive, as the cars ran down the hill themselves while being loaded, the locomotive being employed to haul the empty cars back; the haul was short and a round trip was made in 30 minutes. The report shows that during the work of six months the average number of cars loaded per day was 205, including delays and movings, and that the average cost per cu. yd. was 7 cts., which, as shown, included labor of loading, moving shovel about once a month, moving track to suit, dynamite for caving bank, repairs of shovel, fuel, oil, waste, wages of watchman, rent of cars and locomotives, labor of engineers, firemen and wipers, labor, conductors and brakemen, and, in fact, absolutely everything connected in any way with filling the embankment.

APPENDIX. 59

TABLE IV.

Work Done by Steam Excavator in Six Months at Missouri Valley, Ia.

Repairs to locomotive, shovel and cars; material................	$457.14
Repairs to locomotive, shovel and cars; labor....................	211.80
Supplies for shovel..	1,760.00
Rent of locomotive and cars...................................	1,404.75
Supplies for locomotive.......................................	1,783.52
Wages of locomotive attendants................................	1,508.37
Wages of all other employees..................................	10,680.01
Total cost	$17,803.59
Cars loaded	32,141
Cost per car	55.38 cts.
Cost per cubic yard...	7 "
Hours worked by gang...	2,325
Hours worked by shovel.......................................	1,926

The report of the Roadmasters' Association for 1885 gives the cost of steam shovel work as follows:

Railway.	Work.	Cost per yd.
Baltimore & Ohio..Including everything, haul 5 to 25 miles.......		8.1 cts.
Michigan Central..Loading		4.5 "
Michigan Central..Hauling, 30 miles, labor only..................		4.0 "
N. Y., P., & O....Loading		7.0 "
Central Iowa......Loading	4.75	"
" " .....Unloading.......................................	1.9	"
" " .....Engine service.................................	3.1	"
" " ..Total...	9.75	"

The detailed statement given in Table V. was prepared by Mr. E. A. Hill, Acting Chief Engineer of the Indianapolis, Decatur & Springfield R. R., and is a record of work done under the supervision of Mr. A. J. Diddle, Roadmaster. It shows marked economy and gives an excellent idea of how the expenses are apportioned. The Otis type of excavator was used, which cuts 24 ft. wide and to a depth of 4 ft. below the track. The banks were about 15 ft. high, the average haul 4,000 ft. Twelve flat cars constituted a train. By a special cable arrangement the time of plowing off, ordinarily requiring about 15 minutes, was reduced to 5 or 6 minutes.

TABLE V.

Steam Shovel Work; Indianapolis, Decatur & Springfield R. R.

	Sangamon River Trestle. 1885.	Montezuma Gravel Pit. 1886.	Sangamon River Trestle. 1891.	Guion Trestle. 1887.	Nichol's Hollow Trestle. 1887.
Total number of days......	54	186	48	108	51
Number of working days..	46	115	38	85	40
Days idle besides Sundays.	0	45	3	7	4
Material handled........	light clay.	gravel.	light clay.	light clay.	light clay.
Average height of bank..	10 ft.	12 ft.	10 ft.	10 ft.	12 ft.
Total No. cars loaded.....	2,890	8,631	2,771	5,254	2,528
Greatest No. load. per day	94	124	80	80	75
Least No. cars load. per day	22	16	50	30	15
Average No. loaded per day	63	75	73	61.8	63.2
Average length of haul....	1 mile.	9 miles.	1 mile.	2 miles.	¾ mile.
Grade, shovel to dump, p. c.	—1.00	varying.	—1.00	—1.00	—1.00
Tons coal used,shov. & eng.	141	853	90	170	65
No. car loads per ton coal	20.5	10	28	30.9	38.9

APPENDIX.

Cost of Work Per Car Load.

	Sangamon River Trestle. 1885. Cts.	Monte-zuma Gravel Pit. 1886. Cts.	Sangamon River Trestle. 1886. Cts.	Guion Trestle. 1887. Cts.	Nichol's Hollow Trestle. 1887. Cts.
Foreman at $125 per month	8.86	9.67	8.00	9.01	9.88
Cranesman, $2 to $2.50 day	5.35	5.62	4.80	3.54	5.57
Fireman (shovel) $1.50 day	2.88	3.37	2.87	2.00	3.27
Laborers (4) $1.25 per day	7.86	9.92	8.77	9.80	9.80
Watchman at $1 per day	2.07	1.96	1.88	2.50	2.25
Total shovel crew	27.02	30.54	26.32	27.75	30.77
Engr. and fireman (engine)	12.00	14.50	7.44	11.00	13.10
Trainmen (conductor, $2.50; brakemen, $1.50)	5.97	14.60	5.74	5.25	5.77
Total train crew	17.97	29.10	13.18	16.25	18.87
Helpers distrb. earth, $1.10	....	1.74	....	....	2.72
Sec. men (track work), $1.10	0.81	1.88	1.38	1.45	....
Bridge carpenters (repairs to plant), $2.50	0.15	1.58	0.16	1.04	2.03
Sec. men (reprs plant), $1.10	....	0.62	....	....	....
Shop bills (repairs to plant)	1.69	10.90	1.27	10.00	1.67
Total repairs to plant	1.84	13.10	1.43	11.04	1.67
Coal from $1.25 to $1.41 ton	6.31	13.30	4.47	4.31	3.28
Oil, waste, etc	0.52	1.55	0.75	0.86	0.36
Total supplies	6.83	14.85	5.22	5.17	3.64
Grand total per car load	54.47	91.19	47.53	62.26	59.75
Cost, cu. yd., 8 yds. per car	6.43	11.40	5.94	7.79	7.47
Add " " for interest on cost of plant	1.00	1.00	1.00	1.00	1.00
Cost per cu. yd., includ. int.	7.43	12.40	6.94	8.79	8.47

←—BACON'S—→
HOISTING ENGINES

For Every Possible Duty.

FARREL'S—Ore and Rock
CRUSHERS.

Screens, Elevators, Etc.

THE STANDARD FOR 25 YEARS.

EARLE C. BACON,
ENGINEER,

Havemeyer Building, New York.

WORKS:
PACIFIC IRON WORKS.
FARREL FOUNDRY & MACHINE CO.

Steam Shovel
REPAIRS.

Dippers, Dipper Teeth, Dipper Bails, Dipper Mouths, Kept in Stock.

CHAINS OF ULSTER AND NORWAY IRON.

We have built a new shop, and are preparing designs for

STEAM SHOVELS

of very substantial construction.

We call attention to the fact that we have made for years a specialty of **Pile Drivers** and **Steam Dredges**.

VULCAN IRON WORKS,
CHICAGO.

NEW YORK DREDGING CO.,
ENGINEERS AND CONTRACTORS.

Hydraulic Dredge discharging through 5,700 ft. Pipe. Will dig and put ashore any Material, Rock excepted.

Patent Canal Excavator.

Machines at Work, Washington, D. C.,
Jacksonville, Fla., and Oakland, Cal.

INCORPORATED UNDER LAWS OF NEW YORK.

GEO. W. CATT, M. Am. Soc. C. E.,
President and Engineer.
O. L. WILLIAMS, Secretary and Treasurer.

SPECIALTIES:
Machinery for Economical Excavation of Canals for Dredging; for Reclamation of Low Lands.
CORRESPONDENCE SOLICITED.

WORLD BUILDING, New York, N. Y.

G. L. STEUBNER & CO.,

—MANUFACTURERS OF—

HOISTING BUCKETS

OF ALL KINDS AND FOR ALL PURPOSES.

Side Dumping Cars,
End Dumping Cars,
Bottom Dumping Cars,
Charging Cars,
Special Cars,
Iron Wheelbarrows,
Iron Hoisting Blocks,

Tar Heating Furnaces,
Sheet-Iron Work,
Iron Forgings, Etc.

Send for Catalogue and Price List.

168-176 EAST THIRD ST., LONG ISLAND CITY, N. Y.

THE BEST STEAM SHOVEL CAR.

WRITE FOR PRICES AND CATALOGUE TO

Ryan-McDonald Mfg. Co.,

44 SOUTH STREET, BALTIMORE, MD.,

MANUFACTURERS OF

Light Locomotives, Contractors' Cars, Derrick Irons and Crabs, Hoisting Engines and All Classes of Narrow Gauge Cars.

CONTRACTORS'
...AND...
RAILROAD SUPPLIES.
Construction, • • • •
Dump and Mine Cars.

CATALOGUE AND PRICES ON APPLICATION.

HAROLD C. DAYTON & CO.,
44 DEY STREET, NEW YORK.

M. BEATTY & SONS,

WELLAND, ONT.

Dredges, Ditchers, Derricks and Steam Shovels

OF VARIOUS STYLES AND SIZES TO SUIT ANY WORK.

SUBMARINE ROCK DRILLING MACHINERY,
HOISTING ENGINES, SUSPENSION CABLEWAYS,
HORSE-POWER HOISTERS, GANG STONE-SAWS,
CENTRIFUGAL PUMPS for Water, Sand and
Gold Mining,

AND OTHER CONTRACTORS' PLANT.

Contractor's Locomotives on Hand.

WE keep on hand a number of sizes both narrow and wide gauge locomotives, of best construction, for contractors' service. Catalogue mailed and prices quoted on application.

H. K. PORTER & CO.,
BUILDERS OF LIGHT LOCOMOTIVES,
WOOD STREET, Near 7th Ave., PITTSBURGH, PA.

NO. I SHOVEL ON CHICAGO DRAINAGE CANAL.

OSGOOD DREDGE CO., ALBANY, N. Y.
MFRS. OF DREDGES AND STEAM SHOVELS.

OTIS & CHAPMAN,
STANDARD GRAVEL and **EXCAVATOR**
HARDPAN . . . .

MANUFACTURED EXCLUSIVELY BY

JOHN SOUTHER & CO., Boston.

Earth Displaced at ONE-QUARTER LESS EXPENSE Than by Any Other Machine

To Whom It May Concern:

I hereby certify that I have used the Otis patent improved Steam Excavator the past twenty years, in all kinds of earth excavation, and believe it to be the best dry land excavator in use, and the only one that will work successfully in hardpan material. I have excavated and put into cars five million yards under one contract for making land in Boston; with two of these machines I loaded from seventy to eighty thousand yards per month.

N. C. MUNSON.

J. A. LANE, Manager. **C. P. TREAT,** S. H. DOTY, Engineer.
ROB'T SMITH, Ass't Manager. H. C. DECKER, Cashier

CONTRACTOR BANGOR & AROOSTOOK R. R.

Houlton, Maine, December 31, 1894.

This is to certify that in the month of October, 1894, the bearer, Mr. John B. Shaw, with 1 3-4 yds. Souther Steam Shovel, loaded on cars 38,168 cubic yds. of ballast. Pit measurement by R. R. Co.'s Engineers.

(Signed)   C. P. TREAT,
     per S. H. Doty.

www.ingramcontent.com/pod-product-compliance
Lightning Source LLC
Chambersburg PA
CBHW020252090426
42735CB00010B/1888